FUNGI
Animal, Vegetable or What?

by

David Wright.

Print ISBN 978-1-9192698-9-4

Published by

Llyfrau Cambria Books, Wales, United Kingdom.

Cambria Books and Cambria Stories are imprints of

Cambria Publishing Ltd.

Discover our other books at: www.cambriabooks.co.uk

Acknowledgements.

This book follows earlier published work on how organisms interact with each other and with their natural environment; in particular *"Natural Toxins. The Good, The Bad and The Deadly"* (Wright 2025). I greatly appreciate the help and friendship of my colleague Professor Pamela Welbourn over many years of research and teaching covering several subjects contained herein. I am very grateful for her reviews and discussions of this work and suggestions for potentially useful material. I am much indebted to Sarah Hannis for her invaluable help with chemical structures and other graphics. Also thanks to Anna Counts for several other pictures of fungi.

Above all I thank my wife Susan for her constant support and inspiration throughout.

CONTENTS

INTRODUCTION.

What Are Fungi?

Funguses or fungi are commonly classified as one of the major groups of organisms along with Animalia (animals), Plantae (plants) and Protista. The Protista are a group of unicellular organisms including Protozoa and single-celled members of the algal group Chromista. All these major taxonomic groups, including fungi, are eukaryotes. A eukaryote is a cell or organism that possesses a clearly defined nucleus surrounded by a membrane that also contains hereditary material in the form of chromosomes. This differentiates them from prokaryotes, which are unicellular organisms with no nucleus (e.g. bacteria). Unlike prokaryotes, eukaryotic cells also contain organelles, including mitochondria, which are responsible for energy generation, cell growth and programmed cell death (apoptosis). Fungi include microorganisms such as unicellular yeasts and multicellular moulds, as well as much bigger, multicellular forms such as mushrooms and toadstools.

Fungal Classification.

The taxonomy, i.e. scientific classification, of fungi is extraordinarily complex and has been subject to continuous editing over many decades as new and different taxonomic groups are discovered. The detection of many fungi is often hampered by their cryptic existence in the form of symbiotic[1]

---

[1] A symbiotic relationship between two organisms is a close organic relationship that benefits both organisms. It differs from a parasitic relationship that benefits only the invasive organism, usually to the detriment of the 'host' organism.

or parasitic relationships with other organisms. As new fungi come to light their discovery often leads to the formation of completely new taxa as well as their amalgamation with existing taxonomic groups. Sometimes this leads to the abandonment or reorganisation of older taxonomic groups. Recent advances in biochemistry and genetics have contributed to this process by uncovering hitherto unrecognised phylogenetic relationships that would not have been recognised simply by visual examination.

As an illustration of this complexity, one recent compilation recognises as many as nineteen different phyla as follows:

Aphelidiomycota,	*Kickxellomycota*,
Ascomycota,	*Monoblepharomycota*,
Basidiomycota,	*Mortierellomycota*,
Basidiobolomycota,	*Mucoromycota*,
Blastocladiomycota,	*Neocallimastigomycota*,
Calcarisporiellomycota,	*Olpidiomycota*,
Chytridiomycota,	*Rozellomycota*,
Entomophthoromycota	*Sanchytriomycota*,
Entorrhizomycota,	*Zoopagomycota.*
Glomeromycota,	

Of these fungal phyla the following six are probably the best known and are listed below (Table 1) in more detail:

Table 1. Six Major Phyla of Fungi.

Ascomycota. Often known as sac fungi because of the formation of sexual spores (ascospores) within a sac known as an Ascus. They can reproduce asexually and sexually. They include yeasts, truffles, moulds and edible cone-shaped mushrooms called morels. They are the largest phylum of fungi comprising over 65,000 species including both plant and animal pathogens and may be parasitic or saprophytic[2]. They may also form symbiotic relationships with plants and animals. Several are sources of medicines and therapeutic compounds including perhaps the most famous anti-biotic of all, Penicillin, which is derived from the mould *Penicillium.*
Basidiomycota. Also known as club fungi, members of this phylum produce spores on club-shaped structures known as 'basidia'. They include mushrooms, shelf fungi and puffballs. They are important symbionts of many plants and play an important role as decomposers of plant tissue and other organic debris in woodland areas.
Blastocladiomycota. Inhabitants of both soil and water, these largely aquatic fungi are characterised by the production of flagellated zoospores that are able to swim in water and capable of forming either haploid or diploid colonies. They may live as saprotrophs feeding on dead plant material, although some e.g. *Coelomomyces* are parasitic on (crustacean) copepods and other arthropods such as mosquito larvae.

[2] Saprophytic. Living on dead or decaying matter.

Chytridiomycota (or Caulotrichiomycota), also known as Chytrids, are tiny heterotrophic fungi found in both terrestrial and aquatic habitats. Many are unicellular. They are primarily aquatic organisms that produce flagellated reproductive spores that are able to swim. The spores generated by members of this phylum are characterised by the presence of a single flagellum. However, they also have an amoeboid form that is capable of crawling on solid surfaces. Some are pathogenic on vegetables such as potatoes. Others can cause skin diseases in aquatic and semi-aquatic vertebrates such as amphibians. Chytrids are often seen as ancestral forms along with *Microsporidia*, which has also been described as a fungal phylum, although this remains a subject of conjecture.

Mucoroycota These fungi are generally found in an association with the roots of trees and other plants. They characteristically have symbiotic relationship with the host plant wherein they receive nutrients from the plant tissue in exchange for improving mineral uptake by the plant via the fungal mycorrhizal network[3]. This is a relatively new phylum defined by recent phylogenetic analyses that now incorporate former members of the now-defunct phylum Glomeromycota. These former members of the Glomeromycota are now classified as the subphylum Glomeromycotina in the Mucoromycota phylum.

---

[3] A mycorrhizal network is a symbiotic association between plant roots and a network (mycorrhiza) of fungal strands called hyphae. that perform the same function as roots for the fungus. The joint network is often beneficial to both fungus and host plant.

Neocallimastigomycota. These are small anaerobic fungi found in the digestive tracts of herbivorous animals such as cows, horses and sheep where they aid in the breakdown of plant tissue by the secretion of enzymes capable of cellulitic activity . They are seen as important factors in the decomposition of organic matter and in carbon cycling.

Historic and Current Fungal Groups.

Penicillium.

In April 1928 Scottish bacteriologist Alexander Fleming noticed that *Staphylococcus aureus* bacteria in a Petri dish failed to grow in the presence of the fungal species *Penicillium*, some of which had been accidentally introduced into the dish. Although *Penicillium* had been known since the late 19th century, this was the first discovery of its antibacterial properties. Australian pathologist Howard Florey and British biochemist Ernst Chain isolated and purified penicillin in the late 1930s, a process that led to the development of an effective therapeutic form of the drug by 1941. Fleming, Florey and Chain were awarded the Nobel Prize for medicine and physiology in 1945.

The exact identity of the original penicillin-producing species of *Penicillium* was variously identified as *Penicillium rubrum*, *P. notatum*, and *P. chrysogenum* among others. However, in 2011 phylogenetic analysis confirmed the original source of penicillin as *P. rubens.* Penicillins work by inhibiting the bacterial enzymes that are responsible for cell wall synthesis in replicating microorganisms and by activating other enzymes that break down the protective wall of the microorganism. As a result, they are effective only against microorganisms that are actively replicating and producing cell walls. They therefore do not harm human cells that are not actively engaged in cell division.

Microsporidia.

Microsporidia have been described as a group of spore-forming intracellular parasites of vertebrates, including humans and domestic animals and livestock. They comprise approximately 200 genera and about 1400 species. They are primarily intracellular parasites of the digestive tracts of both vertebrate and invertebrate hosts. In humans, Microsporidia are responsible for several gastrointestinal disorders, now collectively identified as microsporidiosis, which may be particularly serious in immunocompromised patients.

The taxonomy of Microsporidia has been a subject of much debate, with some accounts referring to them as protozoans, others as fungi, with several references remaining uncommitted. However, genetic sequencing evidence (Hirt et al. 1999) indicates that the Microsporidia group belong to the Fungi kingdom.

Slime Moulds. Ex-Fungi?

Slime Moulds were first described by German physician Thomas Panckow in 1654 and were originally grouped in the subdivision Myxomycota, which were historically included in Division Fungi. However, several mycologists now place slime moulds in a separate non-fungal category, Myxogastria, based on the presence of an early developmental stage which has a large amoeba-like[4] form containing several nuclei. Currently the Myxogastria family consists of between 1000 and 1500 species. This amorphous feeding stage grows on damp decaying substrates where it engulfs and feeds upon bacteria.

---

[4] Amoeba-like or amoeboid signifies similarity to an amoeba, which is a unicellular organism capable of changing its shape as it crawls over solid substrates.

Following this growth phase, it dries out and forms structures known as sporangia. These tiny fruiting structures, usually only 1-4mm in diameter develop cellulose cell walls much like plants and divide via meiosis[5] into haploid[6] spores with individual nuclei. Following dispersal these spores land and germinate as haploid cells called swarm cells that combine to form a diploid[7] amoeba. This reproductive strategy, often described as diplontic, is analogous to the meiotic formation of (haploid) male and female gametes in flowering plants (angiosperms), conifers (gymnosperms) and animals and their subsequent union through the process we know as fertilisation. Other organisms displaying a diplontic phase in their development include diatoms and some brown algae. This results in the formation of a diploid zygote, analogous to a fertilised egg, which develops into an embryo, The embryo then undergoes mitotic[8] cell division to form multicellular structures. In several plants the resultant diploid structures are known as sporophytes. The diploid amoebic form that is characteristic of the Myxogastria is seen by some mycologists as equivalent to a sporophyte.

The amoebic cells produced by the Myxogastria have some capacity for movement aided by the presence of two flagella[9]. This further distinguishes them from true fungi, which can develop only one flagellum. These amoebic cells share

---

[5] Meiosis is sexual cell division involving chromosomal division, as opposed to Mitosis (asexual) cell division where the number of chromosomes in the 'daughter' cells remains unaltered.

[6] Haploid refers to a cell or nucleus possessing a single set of chromosomes.

[7] Diploid structures have two sets of chromosomes.

[8] Mitotic (asexual) cell division results in copies that are identical to the parent cell.

[9] A flagellum (plural flagella) is a long thin projection capable of "snake-like" movement enabling a swimming motion. This wavelike action can result in locomotion in both air and water.

some structural similarities with the unicellular blood parasite plasmodium which is responsible for malaria in humans. This resemblance has been responsible for the name 'plasmodial slime moulds' despite the lack of any taxonomic relationship between the two groups of organisms. Having established that slime moulds are not true fungi, the picture is further complicated by the presence of several unrelated or loosely related organisms that display several likenesses to plasmodial slime moulds including the presence of an amoeboid stage and similar fruiting bodies. Examples include Dictyostelia and Protostelia. A good deal of taxonomic uncertainty surrounds these organisms, which have been described as plasmodial slime moulds, but also as distinct sub-groups of amoeba.

The Oomycota, also named water moulds, are a group of plant parasitic microorganisms that, in many respects illustrate the confusion and complications surrounding fungal taxonomy. Oomycetes are classified in a distinct group of fungus-like organisms known as the Stramenopiles that are related to brown alga rather than fungi. This is despite the fact that the word oomycote means egg fungi and the plant diseases caused by oomycete infestation are commonly attributed to the 'oomycete fungus'. *Phytophthora infestans* has been identified as a fungus-like organism known as 'water mould' that is the cause of serious potato and tomato blight. Its current classification includes the clades[10] Pseudofungi as well as Stramenopiles which is a subgroup characterised by the possession of stiff hairs as part of its structure. The oomycetes serve to illustrate that the common names for organisms don't always reflect their true taxonomy.

---

[10] A 'Clade' is a taxonomic grouping derived from the Greek Kládos (class) denoting a common ancestor.

Currently it is estimated that there are between 2 and 5 million species of fungus on earth, although only about 150,000 have been positively identified and described, and the species count is almost certainly subject to further revision. This compares with approximately 300,000 species of plant that have been identified. Until the latter half of the 20th century, fungi were classified in the plant kingdom (subkingdom Cryptogamia) and were separated into four classes: Phycomycetes, Ascomycetes, Basidiomycetes, and Deuteromycetes. The latter are also known as *Fungi Imperfecti* because they lack a sexual cycle. These fungal groupings were primarily defined by the morphology of sexual organs and by the presence or absence of cross walls (septa) in strand-like tissues called hyphae that are an integral part of the fungal organism. However, as our knowledge of the genetics of fungi has advanced, other important identifying markers now include the configuration of nucleic acids and chromosomes in the nuclei of vegetative networks formed from these hyphae. The tissue networks formed from hyphal threads are known as mycelia and are responsible for providing nutrients for the fungi as well as securing their attachment to both organic and inorganic substrates. Mycelia contribute to the nutrition of fungi by secreting enzymes and other chemicals that cause the chemical breakdown of tissues of the biota that are the fungal substrates. Host organisms, mainly plants, may have a symbiotic or parasitic relationship with the fungi they help support. Of the 150,000 or so identified species of fungi, approximately 8000 are detrimental to plant growth and survival. Those pathogenic to humans and other animals number less than 500. In terms of global biomass, plants comprise more than 82% of earth's biota, bacteria 28%, fungi 2.2% and animals only 0.5%.

The Dung Cannon.

The Dung Cannon (*Pilobolus crystallinus*) is a fungus belonging to the phylum Mucoromycota, that grows in clumps on the faeces of herbivorous animals such as cows and sheep, where it aids in the decomposition of the faeces on which it feeds.

The fungus is characterised by a slender stalk between 0.75-1.5" in length supporting a spore capsule known as a sporangium loaded with spores. (FIGURE 1.).

FIGURE 1. **Dung Canon Fungus** *(Pilobolus crystallinus)* inhabits the dung of herbivores and uses an explosive strategy of spore dispersal. Picture by Anna Counts.

The sporangium sits on a highly pressurised vesicle that plays a unique role in propelling the sporangium towards its new herbivore host. This dispersal mechanism relies on sunlight to trigger the explosion of the vesicle which projects the sporangium with a speed measured at 56mph and an acceleration equivalent to 20,000g. This force can throw the sporangium a distance 6.5 feet and is among the fastest biological actions occurring in nature. It is, therefore, not surprising that common names for the fungus also include '*The Shotgun Fungus*' and '*The Hat Thrower*'. The fungal spores continue their dispersal by being consumed by new herbivores. The spores remain viable as they pass through the digestive tracts of grazing animals. Despite its close association with the digestive system of many herbivorous animals, the dung canon is not harmful to consumers.

Another group of (Basidiomycete) fungi sharing the common name 'shotgun fungus' or 'cannonball fungus' belong to the genus *Sphaerobolus*. This genus was first described by the Italian cleric and botanist Pier Antonio Micheli in 1729. Since then, the fungus has undergone a taxonomic reclassification process that currently recognises at least three species, *S. iowensis, S. ingoldii* and *S. stellatus*. Of these, *S. stellatus* is the most common, although the taxonomy of this genus remains in a state of flux. These fungi primarily exist as groups of tiny spherical or ovoid spores housed in cup-like casings 1-3 mm in diameter, which may range in colour from white to orange to brown. These clusters of spore casings, collectively known as peridioles, are found world-wide, growing on wood mulch and other organic debris including faecal matter. The spore casings comprise an inner and an outer layer, best described as two cups, one inside the other.

Like dung cannon, *Sphaerobolus* are capable of discharging spores several feet using a catapult device that is triggered by an extraordinary biochemical mechanism. This involves the transformation of cellular carbohydrates into soluble sugars that result in a quantum increase in the osmotic pressure between the two layers of the spore casing. This rapid increase in osmotic pressure causes both layers of the peridiole to split at their apex. This explosive force is converted into kinetic energy capable of launching spores as the inner casing of the peridiole is everted.

FIGURE 2 depicts a longitudinal section of the sequence of events involved with the slingshot ejection of the spore from its casing. Following the release of the spherical spore the empty peridium assumes a star-shaped appearance (FIGURE 3).

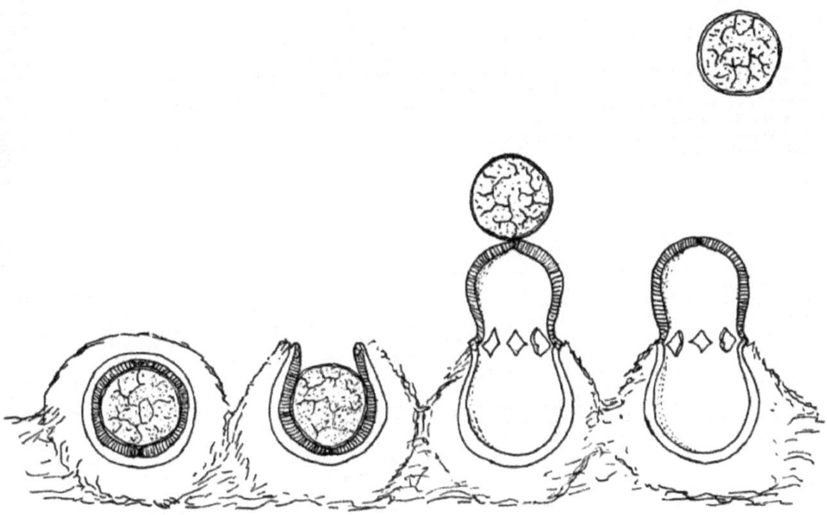

FIGURE 2. Sequence of events involved with the ejection of the spherical spore by the cannonball fungus *Sphaerobolus*

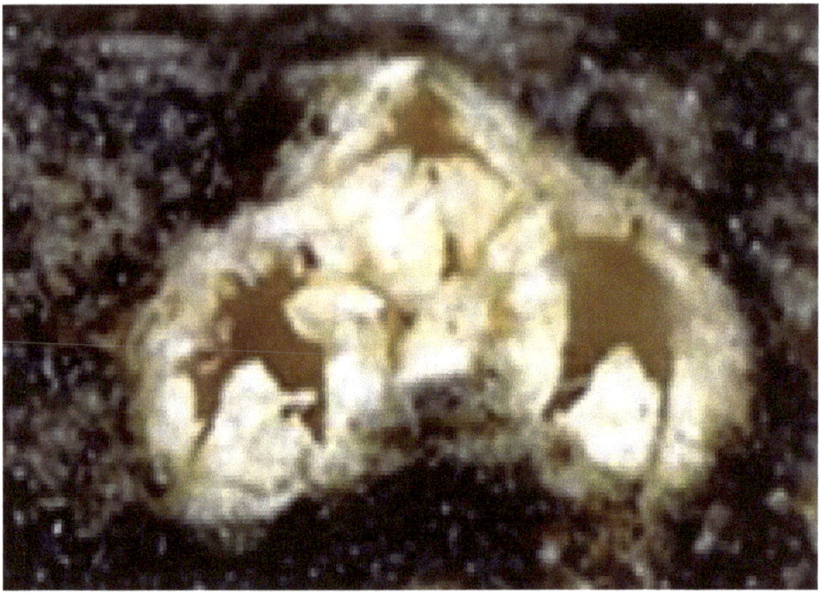

FIGURE 3. Hollow of the cannonball fungus *Sphaerobolus* peridium following spore discharge. As an illustration of the energy involved with *Sphaerobolus* spore dispersal it has been reported that the catapulting phenomenon may be accompanied by an audible 'pop'.

A Humungous Fungus.

Probably the best-known examples of larger fungi are mushrooms and toadstools. Both groups are easily identified by their familiar stalks and caps, known as fruiting bodies, which are of critical importance for fungal reproduction and distribution. Although many fungi are well-known for their cryptic and unobtrusive habitat, some are responsible for mind-boggling statistics related to their biomass and evolution. For example, there is a strong case to be made that certain fungal groups include some of the world's largest living organisms. This somewhat surprising statement can be rationalised when we consider that the only visible form of mushrooms and toadstools are the reproductive structures known as fruiting bodies that appear above ground. In most cases the caps of these fungi rarely exceed 12 cm in diameter, perhaps the size of a small saucer. Many are much smaller.

Habitats for toadstools and mushrooms commonly include woodland terrain, where colonies of these fungi inhabit both the soil and root systems of trees and other plants. The key word here is 'colonies' which, besides the visible and familiar mushroom and toadstool fruiting bodies, include a large subterranean network of thread-like tissues known as hyphae that are woven into a mesh called a mycelium. The mycelium constitutes an integral part of the fungal organism and, as such, forms a major part of the biomass of the total organism. This single subterranean network serves to anchor the fungal colony to the ground or other organisms such as plants, as well as providing a means for absorbing water and nutrients from soil and neighbouring plant tissue.

Although the physical attachment of fungi to plants coupled with the extraction of nutrients from plant tissue may have detrimental effects on host plants there are also several examples of the beneficial effects of fungi on neighbouring plant-life. In such cases, fungi may form symbiotic relationships with plant hosts in which mycotoxins secreted by invasive fungi act as deterrents of potential consumers of both embedded fungi as well as their host plants. In other words, fungi may extend their protective properties to plants they have colonised. Additionally, fungi have been shown to improve the water retention capability of plants with which they are partnered, an important asset in a dry climate. Another hyphal derivative performing these functions in some species of mushrooms and toadstools has been given the name rhizomorph. Like mycelia, rhizomorphs form networks that function as an integral part of the fungal organism. As with mycelia, rhizomorphs attach the fungi to their respective substrates and are capable of extracting nutrients from host organisms such as tree bark and from other organic forest floor debris that contributes to the fungal habitat. However, unlike mycelia, which are usually white or pale in colour, rhizomorphs usually take the form of black cordlike structures, which has earned them names such as ‘shoestring’ or ‘bootlace’ fungus in some published work. The hybrid term ‘rhizomorph mycelium’ has also been used to describe these root-like complexes, which are also an integral part of the mushroom and toadstool colonies where they are found.

In view of the enormous scale of some of these fungal colonies and the role of the rhizomorph system or mycorrhiza in expanding their spread it is reasonable to view this root system as a means of fungal mobility resulting in the colonisation of extensive areas of terrain. However, their distribution is also aided by the release of reproductive spores, some of which may be flagellated to aid their dispersal. Spores may be widely dispersed via air, water, or by attachment to other organisms including animals which may serve as vectors for a wide variety of fungi. Spores are capable of remaining dormant during much of the time spent free-living in the natural environment, thereby withstanding potentially damaging environmental conditions and ensuring their survival, shelter and distribution pending their association with a substrate that supports their colonisation and propagation. Many adopt parasitic or symbiotic relationships within the tissues of other organisms, thus providing further protection from adverse weather and other environmental conditions.

A famous example of a mushroom possessing a rhizomorph system is the Honey Mushroom (*Armillaria ostoyae*), which has the mind-boggling distinction of being ranked as the worlds' largest single living organism. An example of the physical scale of a specimen of this species was illustrated by a 1998 study of a Honey Mushroom colony from Malheur National Forest located in a mountainous region of Oregon, U.S.A. This single colony was confirmed by tissue DNA analysis to represent a solitary specimen and was measured to cover areas as large as 10 square kilometres. Estimates of the total biomass attributed to this specific Honey Mushroom fungus have varied between 6000 and as much as 35,000 tons. With respect to biomass, it is interesting to consider that, by comparison, a single specimen of the world's largest-ever animal, the blue whale (*Balaenopterus musculus*), weighs in between 100 and 200 tons!

Decision-making Fungi?

Because of the subterranean existence of most fungal mycelia the chief indicators of the presence of fungi are their fruiting bodies such as mushrooms that protrude above the ground. Although the vast majority of mycelia are subterranean, furry outgrowths of mycelia may be spotted on woodland floor debris such as fallen branches and leaves. These mycelial outcrops signify the presence of foraging targets for fungi in their quest for organic food material. The words 'quest' and 'foraging' are used deliberately here to indicate an apparent degree of intent on the part of these mycelia; a behavioural habit suggesting that some fungi seem to have an 'awareness' of their surroundings and the ability to react to certain stimuli such as the presence of food.

It may be argued that this is no different from plants spreading their roots to take advantage of the presence of water and nutrients in the soil. However, several fungal mycelia appear to take this at least one step further. Examples include their apparent ability to actively forage for nutrients and their complex behaviour traits that ensure maximum deployment of resources to colonise food items such as leaves and pieces of wood. This approach is best explained through direct comparison between mature plant roots and fungal mycelia.

Plant root systems tend to adopt a 'blanket approach' to obtaining nutriment, although 'trial and error' might be a better term. Parts of a plant's root system that do not encounter water and nutrients will atrophy and die, a process that may take several years. In contrast, a fungal mycelium will respond to the presence of a nutrient source within a matter of days, aborting any hyphal branches that return no reward in the form of potential food. Instead, they propagate and extend fresh tendrils towards unexplored territory in a renewed quest for nutrition.

Information obtained by exploratory hyphal filaments on the presence or absence of potential food sources appears to be transmitted throughout the whole mycelial complex, which then reacts as a single entity to maximise its resources in order to obtain that food. Observations of foraging mycelia have, therefore, demonstrated that mycelia do not waste energy sending multiple tendrils to barren areas but will instead send extra filaments to locations where hyphae have located new potential sources of nutrients.

FIGURE 2 shows the results of an experiment by mycologists showing the migration of a mycelium belonging to the wood-rotting fungus *Phanerochaete velutina* towards a wooden block placed next to the fungus. Once the food source (in this case a block of wood) is located by exploratory tendrils, the radial expansion of the mycelium is abandoned in favour of vigorous and highly directed invasion of the target wood block. Of particular note is the abandonment and excision of hyphae not engaged in colonising the new food source

FIGURE 2. **Phanerochaete velutina**. Foraging between two wooden blocks over a 48-day period. Photos taken by Yu Fukasawa.

In a more elaborate experimental design Fukasawa and co-workers were able to demonstrate a form of pattern recognition by this wood-rotting fungus. In this experiment, a total of 9 wooden squares impregnated with mycelial extracts of the fungus *Phanerochaete vetulina* were arranged in a figure O or an X pattern (FIGURE 3.). Following an incubation time of up to 113 days, mycelial growth was documented in terms of hyphal 'connectedness' (number of filaments) and hyphal mass on each of the wooden squares, which was recorded as a function of the weight loss due to its consumption by the fungus. A feature of the experimental design was that the wooden squares arranged in the circular configuration shared a uniform topography relative to each other. However, the cross-shaped configuration resulted in uneven distancing among the colonised squares depending on whether they were closer to the centre of the cross or on its perimeter.

The most extraordinary finding of the experiment was that the fungus was able to adjust its connectivity and growth patterns in ways that took into account the overall shape of what might be termed the 'super-colony' (i.e. circle or cross). These adjustments in growth and connectivity were made in a manner indicating that the mycelial complex could recognize shapes and was able to communicate information about its surroundings throughout its entire network and adapt accordingly. Inevitably this study has prompted much discussion about whether this fungal behaviour was a product of what might described as 'intelligence'. If we concur with the dictionary definition of intelligence as 'the ability to gain and apply knowledge and skills', then we come close to agreeing with this definition. If so, then the question becomes one of sentience.

In the words of the Chief Scientist who conducted these experiments, Yu Fukasawa:

"Our result on fungal intelligence may help people to notice that many part[s] of intelligent behaviour of organisms can be achieved without brain and consciousness,"

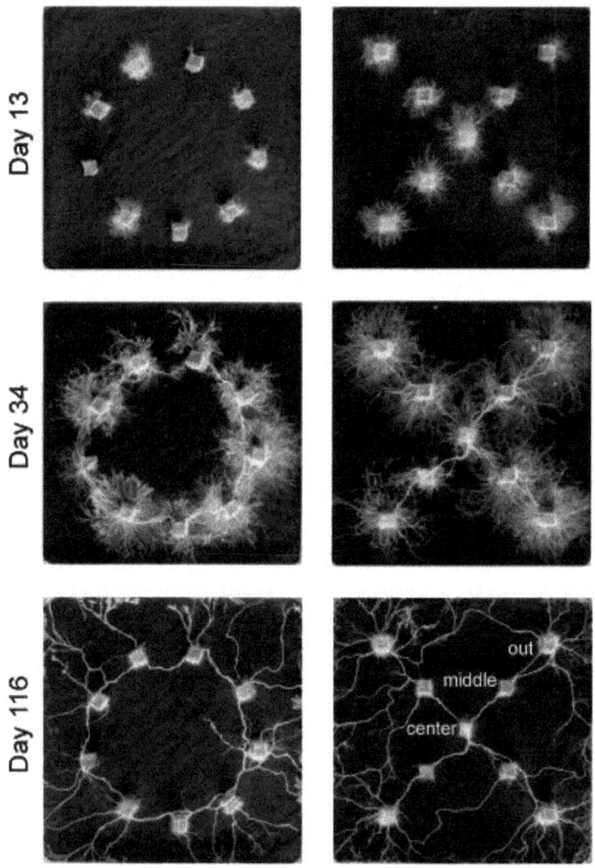

FIGURE 3. Mycelial network of the wood-rotting fungus (***Phanerochaete velutina***) grown on wooden blocks (Fukasawa et al. 2024).

The Mystery of Fungal History.

Using information on the rate of spread of Honey Mushrooms at the Oregon location, calculations have also been made of the age of the overall colony, which has been conservatively estimated as 2,400 years. However, some estimates place this and other North American mushroom networks as much older in origin, some as old as 6000 years! Based on nucleic acid analysis and other genetic markers it has been confirmed that these mushroom colonies can be considered as single organisms rather than assemblages of conspecific or closely related biota. This evidence indicates that these are indeed truly ancient <u>individual</u> organisms. Bearing in mind that *Armillaria* colonies are essentially single organisms, the nickname for this genus "humungous fungus" is well earned. Other species of *Armillaria,* usually described as variants of Honey Mushroom, are also numbered among the largest living organisms known to have existed, although few can compete with the Oregon colony mentioned above. For example, a specimen of Honey Fungus from Crystal Falls Michigan, identified by some researchers as the original "humungous fungus", has now been estimated to be a "mere" 400 tons (although both Oregon and Michigan colonies seem comparable in age).

Based simply on plant-like morphological characteristics of fungal organisms, the early study of fungi, known as Mycology[11], was originally regarded as a branch of botany. However, it is now believed that fungi are genetically more closely related to animals than to plants, although their evolution and taxonomic relationships remain topics of much debate. One clue that places fungi closer to animals relates to the fact that, like animals, fungi respire oxygen and give off

[11] Mycology, the study of fungi, is derived from the Greek μύκης, *mykes* 'mushroom'.

carbon dioxide (CO_2), whereas fungi, in common with animals, 'inhale' oxygen and produce CO_2 as a waste product.

Part of the uncertainty of identifying and dating fungi relates to the huge diversity in their physical characteristics and their symbiotic or parasitic relationship with plants and animals which may hide or disguise their presence in the environment. In parasitic or symbiotic mode many fungi may become deeply embedded in their host organism and in several instances, they may form hybrid products known as lichens in combination with cyanobacteria and/or unicellular algae. In such cases the algal and fungal components of the lichens adopt a symbiotic relationship.

From an evolutionary standpoint the time course of migration of fungi from the aquatic to the terrestrial environment remains the subject of speculation, although it is generally assumed that the transition of fungi from water to land began in the Cambrian era more than 500 million years ago, with evidence of fungi commonly appearing in the fossil record from the Devonian period 400 million years ago. Fungi have since undergone several other evolutionary peaks. For example, during a Cretaceous-tertiary extinction event approximately 66 million years ago, most animal life on earth was wiped out following collision of a massive meteorite with the planet. This event together with subsequent volcanic activity resulted in the disappearance of 80% of animal life on earth. This mass extinction, which effectively marked the elimination of the dinosaurs, was caused by both chronic atmospheric pollution exacerbated by major volcanic activity and an extended era almost totally devoid of sunlight. Animals obliterated by the extinction event ranged from small planktonic organisms to large reptiles, both terrestrial and aquatic. On the other hand, the cryptic, opportunistic habit of fungi enabled them to thrive by living on either living or dead biota throughout this period. It is noteworthy that the

Cretaceous-Tertiary extinction was one of five mass extinction events over the last 500 million years that have resulted from a variety of planetary and climactic events.

In total, these are:

Ordovician-Silurian. 440 million years ago.

Devonian. 360-410 years ago.

Permian. 252-265 million years ago.

Late Triassic. Ca. 200 million years ago.

Cretaceous-Tertiary. Ca. 66 million years ago.

Of these mass extinctions, the Permian event was probably the most damaging, responsible for the destruction of approximately 70% of the Earth's land animals and over 95% of its marine life. In all of the extinction cases shown above there have been some recorded signs that fungi were less affected by these events than other life forms. Evidence for a consequent boost in fungal biomass associated with these extinction events again points to the ability of fungi to permeate and successfully feed on live or dead host organisms, thereby giving them an advantage over other life forms in a hostile climatic environment.

Some observers have speculated that the planet earth has already embarked on a sixth extinction event triggered by man-made climate change. A logical conclusion from such an outcome would be that humans may be inadvertently helping provide a future environment that could help fungi to prosper. A corollary of this is, of course, that humans themselves would be part of this extinction event. Only time will tell; and that will take several million more years!

Compared with plants and animals there is only a relatively sparse fossil record for fungi and, in recent decades, evidence for their evolution has increasingly relied on aspects of their genetic profile to clarify their ancestry and taxonomic affiliations. For example, evidence from fungal DNA analysis has indicated their descent from a flagellated aquatic antecedent between 1.3 – 1.5 billion years ago. However, in 2017, Bengtsen et al. reported the discovery of organisms resembling fungi in much older igneous rock derived from West Griqualand, which is now part of North Cape Province in South Africa. The basaltic rock from this area is commonly referred to as the Ongeluk-Hekpoort magma, dating back as far as the Paleoproterozoic era 2.5 billion years ago. This discovery suggested that the fungi embedded in this rock may redefine fungal origins to an era 2-3 times older than originally thought. Many of these organisms embedded in the Ongeluk rock were characterised by the presence of filamentous strands, called hyphae, that were capable of fusing into the network now known as the mycelium. The mycelium is widely considered to be the precursor to the root system of terrestrial plants, and the symbiotic relationship between fungal mycelia and green algae is widely seen as a key event in the transition of plants from the aquatic to the terrestrial environment. The structure formed from the symbiotic association between fungal mycelia and algal tissue is known as the mycorrhiza, which has evolved into the root system of most terrestrial plants. Today, more than 90% of all plants rely on this tissue for the absorption of water and nutrients from their immediate terrestrial environment. (Wright 2025). The large surface area of the root system derived from the mycorrhizal network is an essential requirement for this vital function,

Apart from lichens, which often have their own unique physical characteristics, a variety of fungal forms may become established within different tissues of host organisms, with either beneficial or detrimental effects.

A characteristic that places fungi in a different kingdom from plants, animals and unicellular organisms such as protozoa, is the presence of the polysaccharide[12] chitin as a constituent of their cell walls. Fungi, like animals, are heterotrophs; they are unable to photosynthesize and acquire their food by absorbing dissolved molecules, typically by secreting digestive enzymes into their immediate environment. Fungi are often seen as opportunistic, relatively harmless inhabitants of other biota with the capability of absorbing nutrients from dead and living tissues of other biota. However, many fungi are by no means benign!!

As well as assisting in the decomposition of dead plant tissue parasitic fungi can often cause the death of host plants. In some cases, fungal activity can result in the wholesale destruction of large areas of woodland. The previously mentioned honey mushroom *Armillaria* and other members of the same genus are responsible for the deaths of whole swathes of forest pines and other trees. *Armillaria* root rot is a common disease of North American conifer forests infested with honey mushrooms.

The Blister Rust.

Several other fungi such as rusts have also been responsible for the rapid deaths of some species of North American conifers. A fungal species of particular concern has been the blister rust *Cronartium ribicola* which seems to specifically target the White Pine *Pinus strobus* and the Western White Pine *Pinus monticola.* Primary symptoms of Blister Rust

[12] Polysaccharides are long chains of carbohydrate molecules that are composed of smaller monosaccharides. Monosaccharides are the simplest form of carbohydrates and are known as sugars.

disease are the development of blisters on the bark of affected trees and the browning and shedding of their needle-like foliage. The Blister Rust fungus was introduced into North America from its native China and other Asian countries in the early 1900s and has since been a subject of critical concern for the North American forestry industry. A feature of the disease is its somewhat variable pathology within the *Pinus* genus. For example, the Western White Pine appears to be particularly vulnerable to Blister Rust. However, some Asian White Pines seem quite resistant to the disease. For example, the Macedonian White Pine demonstrates good immunity to the Blister Rust disease. This resistance may exhibit a genetic trait stemming from a long evolutionary history of co-existence of potential hosts (trees) and disease organisms (rusts), i.e. antagonistic host/pathogen reaction(s) may have taken second place to shared adaptation to a common environment over many generations.

In many cases the relationship between fungi and host plants goes well beyond what might be termed 'benign co-tolerance'. Indeed, despite well-publicised, notorious examples of the detrimental effect(s) of fungi on trees and other plants there are numerous reports of instances of reciprocal interactions that benefit both fungi and their host plants. Indeed, in some cases it has been shown that the very survival of selected plants has relied on their association with certain species of fungus.

The importance of the forestry industry has given rise to a massive literature concerning the importance of what is often called the ectomycorrhizal (ECM)[13] environment. Much of this

---

[13] An ectomycorrhiza is a symbiotic partnership between a fungus and the root system of a plant that benefits both the fungus and the host plant. Such a partnership may provide nutrients as well as mutual protection against

work concerns the taxonomic (and often genetic) make-up of forest fungal communities and how to best pair them with commercially important tree species that benefit from the established fungal community. Emphasis is given to the propagation and preservation of a fungal (ECM) environment that is best suited to support what is called 'sustainable forest management'. In other words, the primary focus for much of this research is finding a ECM environment that helps to sustain a symbiotic relationship with newly introduced commercially important trees. Although the details of these papers and reviews are beyond the scope of this book, typical lines of research follow attempts to establish effective symbiotic fungal/tree partnerships that improve the supply of nutrients to commercially important species of tree. For example, this may involve the establishment of specific tree species in an ECM milieu capable of providing a supply of plant nutrients in an otherwise nutrient-poor environment, or where a cold climate may limit the ambient nutrient supply.

Arbuscular or endo- mycorrhizal fungi are another fungal group capable of forming a symbiotic relationship between their subterranean mycorrhizal networks and numerous plant root systems. Although arbuscular mycorrhiza are confined to fungi of the Glomeroycota group they are found in more than 80% of plant families where they take the form of intracellular networks called arbuscles within the cortical cells of the host plants' root system. Like ectomycorrhiza these facilitate the interchange of plant-based phosphate, sulphate and nitrate accumulated by the fungus with carbon-based nutrients derived from the plant. This mutually beneficial trade of essential minerals and nutrients between the fungus and the plant root is aided by the 100- to 1000-fold increase in the surface area created by the presence of the mycelium. In other

---

mutually common consumers. The ECM environment describes the habitat surrounding the immediate fungal/root environment.

words the arbuscles act as extensions of the root system of the plant. In exchange for providing the plant greater access to important N, P and S-based nutrients from the soil the fungus receives from the plant carbohydrates and other products that they are incapable of producing themselves because of their lack of photosynthetic capability.

Fungi are particularly noted for their ability to facilitate the organic breakdown of other organisms and tissues through the release of enzymes and other digestive agents secreted by fungal mycelia. This digestive activity results in the breakdown and release of chemical by-products that can be assimilated and used by neighbouring biota, including fungi themselves acting in parasitic mode. In this role they tend to act as 'decomposers' in ecological systems. These and other physical and molecular characteristics place fungi in a single group of related organisms, named the *Eumycota* (true fungi or *Eumycetes*), that share a common ancestor (i.e. they form a monophyletic *group*). This interpretation is also strongly supported by molecular genetics. This fungal group is distinct from the structurally similar myxomycetes (slime moulds) and oomycetes (water moulds).

Fungal Evolution. How Old is Old?

A review of current literature concerning the history and taxonomy of fungi illustrates significant remaining areas of uncertainty, and it is clear that the evolution of these organisms is a story that is still being told. For example, the Microsporidia, a group of spore-forming unicellular parasites that were once assumed to be protozoans or protists, are now considered to be true fungi, or a sister group to true fungi.

The discipline of biology devoted to the study of fungi is known as 'Mycology', which is derived from the Greek for mushroom, μύκης, *mykes*. In the past, mycology was regarded as a branch of botany, although it is now known that fungi are genetically more closely related to animals than to plants. A corollary of this is that the ancestor from which fungi diverged over a billion years ago was shared with the animal kingdom as opposed to the plant kingdom. Members of this ancient common heritage group or clade, including forerunners of the both the Fungi and Animals, have been given the name 'Opisthokonta' which is derived from the ancient Greek 'opisthios' meaning 'rear' and 'kontós "pole" or flagellum. In other words, this group of organisms is defined by the presence of a (single) posterior flagellum that is responsible for their forward propulsion. Perhaps the most similar uniflagellate image that might be encountered today would be the mammalian sperm cell. Two other flagellated sister forms with protozoan characteristics, 'Breviata' and 'Apusomonadida' are also part of this group, and, together with Opisthokonta, comprise what has been referred to as the "super-clade" Obazoa.

In summary, it is concluded that the forerunners of fungi and animals together with several flagellated/amoeboid unicellular relatives are represented by the three groups Apusomonadida, Breviata and Opisthokonta, and that, together, these groups constitute the ancestral clade Obazoa. As we have seen, some accounts have assumed that the divergence of fungi and animals from their evolutionary ancestor(s) was a process that occurred 'more than one billion years ago'. Our earlier reference to genetic evidence quotes a transitional period 1.3 – 1.5 billion years ago. However, any figure that we might arrive at must, eventually, be reconciled with the observations of Bengtson et al. (2017), who reported the presence of fossilized 'fungi-like' mycelia in 2.4-billion-year-old volcanic rock from a massive igneous area comprising much of the current state of South Africa and several surrounding countries. We are, therefore, confronted by an apparent 'discrepancy' of <u>one billion years</u> in the appearance of at least two major phyla. On one hand we can use DNA information to place the emergence of fungi (and animals) from their forerunner(s) approximately 1.4 billion years ago. On the other hand we see fossil evidence of the existence of fungi in volcanic rock that apparently pre-dates that transitional estimate by as much as one billion years! It is tempting to use the phrase 'give or take one billion years' in defining fungal evolution, and we are reminded that some areas of science can be somewhat inexact! Others, however, are quite precise, and when it comes to exploring and quantifying the interactions of fungi with other organisms over recent millenia, the level of precision can rapidly reach the molecular level.

Mycotoxins.

The story of parasitism and defensive action by potential host organisms has often been described as biological warfare. In this regard competing plants and animals are often cast in the role(s) of attacker or defender, in which they deploy a broad range of natural toxins to neutralise or overcome their adversaries and competitors (Wright 2025). Fungi are often no different in this respect, and the term mycotoxins is used to describe the bioactive compounds they produce. Like other natural toxins these chemicals may have positive or negative effects on other wildlife. Some examples of biological warfare involving fungi are described in the following examples.

Fungal Infestation. A story of offense and defense

A Smutty Battle.

Ustilago (or *Myosarcoma) mardis*. Is a pathological plant parasitic fungus which attacks maize and teosinte, an important Mexican cash crop related to corn. The resulting pathology is known as Corn Smut, a name shared by the fungus itself.

The affected plants have evolved robust defence systems against pathogenic microbes such as *U. maydis*. A defensive reaction in response to a pathogenic fungal attack may take the form of an 'oxidative burst' where the plant produces reactive oxygen species at the site of the invasion by the fungus. An oxidative burst is typically triggered by the activation of the enzyme NADPH oxidase[14], and results in the generation of Reactive Oxygen Species (ROS). ROS are unstable molecules with strong oxidative properties which have a damaging effect on macromolecules such as proteins, nucleic acids and lipids. These disruptive cellular effects can cause physical damage to membrane integrity and numerous tissue functions, and represent a powerful antimicrobial defence against a pathogen such as Corn Smut (*U. maydis)*. However, the fungus can respond to the host's defensive oxidative burst with its own counteractive oxidative stress response regulated by the gene YAP1. This example of biological warfare response protects *U. maydis* from the host defence and reinforces the pathogen's virulence. Additionally, the fungus has a well-established recombinational DNA repair system which acts during sexual cell division (meiosis) and

[14] NADPH oxidase is nicotinamide adenine dinucleotide phosphate oxidase.

asexual cell division (mitosis). This helps the Corn Smut survive DNA damage arising from the host plant's oxidative defensive response. Despite the defensive measures, corn smut infection nevertheless results in large outgrowths that engulf the kernels of the host plant and result in the development of large bulbous growths, often referred to as galls, that displace the corn cobs.

An interesting consequence of Corn Smut parasitism of the Mexican corn plant, teosinte, is that the galls produced as a result of fungal infestation have become a much-sought-after (and expensive!) culinary delicacy known as huitelacoche. These bulbous growths which consist of a mixture of corn and Smut Fungus are credited with a pleasant creamy flavour as well as having antioxidant, antimicrobial and anti-inflammatory qualities.

Truffles. Food of the Gods.

As a result of their reputation as highly prized edible delicacies, huitelacoche have often shared a culinary niche with truffles, many of which command exceptionally high prices, despite the fact that these fungal groups have little in common apart from their fungal grouping. Perhaps the most expensive of is white truffle (*Tuber magnatum),* which may command prices as a high as $9000 per pound.

Truffles can best be described as underground mushrooms and are, in fact the subterranean fruiting bodies of fungi belonging to the Division Ascomycota, a group known as sac fungi. The term sac fungi refers to their characteristic of carrying their reproductive spores in an enclosed sac-like structure.

As many as 100 genera of fungi have been classified as truffles, including members of the genera *Geospora,*

Choiromyces, Leucangium, Pezizo and *Tuber* within the class Pezizomycetes. It seems somewhat confusing that many truffles are currently classified in the genus *Tuber* within the fungal class (Pezezomycetes), because the word 'tuber' is of course well-known as a descriptor of a group of vegetables, notably yams and potatoes that have nothing to do with fungi. The common usage of the term 'tuber' simply reflects the swollen, rounded nature of the truffle's fruiting body and the swollen root tubers that are characteristic of yams and potatoes.

Although truffles are subterranean, they occupy a forest floor habitat similar to many mushroom colonies and, like mushrooms, they are commonly located growing in close association with trees. Truffles are often found close to beech hazel and oak, although they have also been shown to cohabit with poplar, birch and pine trees. The Black Truffle ***Tuber melanosporum.*** (FIGURE 4.) Is a subterranean species found in Europe and highly prized in the Perigord region of Southwest France.

FIGURE 4. The Black Truffle ***Tuber melanosporum.***

The history of the consumption of truffles by humans probably goes back as far as 5000 years with records of their extraction and preparation as food appearing first in Babylonian texts and, later, Sumerian texts as early as 3000 BCE. Truffles also feature in the collection of Roman cookery recipes entitled *De re Coquinaria*, that have been attributed to a Roman cook named Apicinus. However, the provenance of this author remains uncertain, and the French historian Bruno Lauroux, an expert on culinary history, has suggested that several of these Roman recipes may post-date the imperial Roman era and that most or all of these recipes might belong to the fifth century or later. For further clarification, interested readers should consider John Edwards' book *The Roman Cookery of Apicius* (Edwards 1984).

One reference to the appetising nature of truffles that falls squarely within the Roman imperial age is reputed to have come from the emperor Nero who reigned from AD54 – AD68. In this quote Nero is reputed to have described truffles as "The food of the gods". However, the degree to which this quote can be taken 'at face value' has been questioned by those who attach some scepticism to its true meaning considering Nero's own history with fungi.

It might be remembered that Nero was the beneficiary of the death in AD54 of the emperor Claudius at the hands Nero's mother Agrippina. It was Agrippina who supplied Claudius with the poisonous mushrooms that killed him. She then encouraged, some might say arranged, the marriage of her son Nero to Claudius' daughter, thereby ensuring Nero's accession to the position of emperor. Considering the close taxonomic relationship between mushrooms on one hand and truffles on the other, a sceptic might conflate the benefits to Nero of the two fungi. In other words, when using the phrase 'food of the gods', was he thinking of delicious truffles or was he thinking of the mushrooms that assured his rise to power?

In addition to their culinary history truffles also had a reputation in ancient Greek and Roman folklore as an aphrodisiac beginning with writings of the Greek philosopher Pythagoras, who clearly had other things on his mind besides the geometry of right-angled triangles!

The mystical properties of truffles appeared in the *Naturalis Historia* (77-79AD) written by the Latin scholar Pliny the Elder, who considered the truffle 'A Miraculous Product of Nature'. During the same era the Greek philosopher Plutarch of Chaeronia elaborated on the mythological qualities of truffles, claiming them to be a product of a lightning bolt hurled by the Greek God Jupiter (synonymous with the Roman god Zeus) close to an oak tree. This story was also popularised by the Roman poet Juvenal, who supported the idea that truffles were products of Jupiter (Zeus) comprising lightning, thunder and rain. The aphrodisiac properties of truffles were also elaborated by Roman-Greek physician Galen to whom is attributed the somewhat elegant belief that truffles "--- produce a general excitement that predisposes to voluptuousness".

Caterpillars and Mushrooms.

An Unlikely Pairing that is Stranger than Fiction? Maybe not!

Those of us familiar with Lewis Carrol's story of "Alice's Adventures in Wonderland" will be aware of young Alice's ventures into a fantastical world characterised by dramatic shifts in size and perspective and the altered perception of time and space. Following Carrol's shape-shifting world full of unlikely pairings and bizarre relationships, several readers have wondered whether some of the symptoms exhibited by Alice bear some resemblance to the hallucinogenic syndrome desynchronization, and that the author Lewis Carroll may have indulged in the consumption of some psychoactive potions to gain access to his fictional 'Wonderland'. In other words, it is speculated that Carroll himself might have followed his fictional heroine's disregard of the "Do Not Drink" (the Magic Potion) notice that helped her on her journey.

However, any involvement of Magic Mushrooms or other mind-altering substances in the creative process takes a hit when we remember that Lewis Carrol is really the poet, mathematician and Anglican Deacon Charles Lutwidge Dodgson. Dodgson's profile seems a long way from a psychoactive practitioner and closer to a very good children's storyteller. Dodgson was ably assisted in his work by the excellent illustrator John Tenniel who was responsible for over 40 illustrations accompanying Dodgson's *'Alice In Wonderland'* narrative. Among the strangest of these is a hookah-smoking caterpillar conversing with Alice while relaxing atop a large mushroom (FIGURE 5). 'Surely another sign of hallucinogenic indulgence involving mushrooms' it has been argued. However, like Charles Dodgson, Tenniel's 'pedigree' appears to be above reproach. After all Sir John Tenniel was chief political cartoonist of Punch magazine at that time of his career.

Although there is not a scrap of evidence to link magic mushroom consumption to the authorship and illustration of *Alice in Wonderland*, back in the real world we find a surreal relationship between caterpillars and mushrooms that has many of the ingredients you might expect in fantasyland. Evidence, perhaps, that truth is sometimes stranger than fiction!

FIGURE 5. John Tenniel's illustration of **Alice in Wonderland** being confronted by a hookah-smoking caterpillar sitting on a (magic?) mushroom.

Caterpillar Fungus. Himalayan Viagra (and Gold).

A parasitic fungus with highly prized spiritual and aphrodisiac properties is the Caterpillar Fungus *Ophiocordyceps sinensis* which is a product of eastern Asia, notably Tibet and China, where its mystical and medicinal qualities have long been part of the folklore of those countries.

The fungus infests caterpillars of so-called ghost moths, which are members of the Family Hepialidae (Order Lepidoptera) native to the Himalayan regions of India and Tibet. *O. sinensis* infestation of the moth larva occurs as it feeds on subterranean tissues of other plants. After invading the caterpillar, the fungus consumes the larva from the inside, causing its mummification and eventual death.

During the infestation process the fungus produces a dark brown or orange stalk-like fruiting body containing a mycelium that spreads throughout the body of the caterpillar. This culminates in the death of the caterpillar and its complete colonization, with the fruiting body of the fungus protruding from the head of the dead caterpillar. This bizarre combination now resembles the 'medicinal mushroom' that was first recorded in the 15th century by the Tibetan doctor Zurkhar Namnyi Dorje. Its Tibetan name yartsa gunbu translates as "summer grass winter worm", a phrase that reflects its appearance as a stick-like protuberance from the wormlike body of the caterpillar (FIGURE 6.).

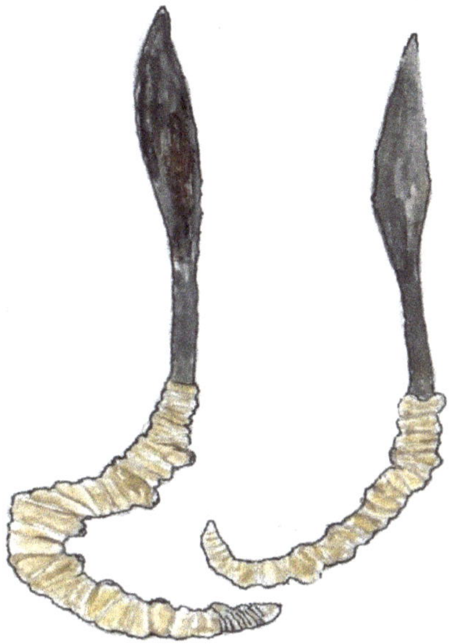

FIGURE 6. **The Caterpillar Fungus** (Ophiocordtceps) is a genus of fungi that infects larval insects, notably the caterpillar of the ghost moth native to Eastern Asia.

The species was first given the taxonomic name *Sphaeria sinensis* in 1843 by the botanist and clergyman Miles Berkeley but was renamed *Cordyceps sinensis* in 1878. This name persisted until 2007, when updated molecular analyses of the fungi resulted in the transfer of several *Cordyceps* species, including *Cordyceps sinensis* to the genus *Ophiocordyceps*. The caterpillar fungus is, therefore currently known as *Ophiocordyceps sinensis* (FIGURE 6.) Nevertheless, other species names have been commonly used to describe the caterpillar fungus, and it seems probable that several organisms can be regarded as synonymous with *O. sinensis*. One example is *Cordyceps militaris*, which has also been given the name caterpillar fungus. This fungus was originally described as *Clavaria militaris* in 1753 by Carl Linnaeus but was renamed *Cordyceps militaris* in 1818. The fungus has since been given at least 8 different taxonomic identities. This probably reflects the parallel evolution of multiple conspecific or closely related organisms occupying a highly characteristic niche in the fungal world.

The cold Himalayan origin of the caterpillar fungus and the resulting characteristically slow development of this fungal/insect duo have meant that the fungus remains a very rare natural phenomenon that is obtained almost exclusively by harvesting it from its natural environment, a lucrative activity for the local populace. *The Kathmandu Post* reports that harvesting the caterpillar fungus can net a Nepalese salary which is higher than the average per capita income from other activities. Each year more than 150,000 individuals are engaged in harvesting the fungus during the May/June season. Usually, however, it is the traders that make more money than collectors.

Attempts to cultivate *Cordyceps* fungi and caterpillars together on a commercial basis have, thus far, met with limited success, although the high market value attached to the fungus and its caterpillar host still provides a powerful incentive for its cultivation, not to mention some fraudulent claims associated with bogus 'caterpillar fungi'. The rarity and exotic nature of the product coupled with its much-sought-after remedial properties all contribute to its extraordinarily high demand and market price. Driven by a vigorous black market, the current price of the caterpillar fungus has been known to reach and sometimes exceed prices that are more than twice the market of the white truffle. Quite a sum for a mummified caterpillar!

In addition to its reputation as an aphrodisiac according to Chinese folklore going back more than 1000 years, the caterpillar fungus has been credited with an impressive array of medicinal properties including the treatment of numerous heart, liver, intestinal, respiratory and renal diseases. Many of these remedies are believed to result from the ability of the active ingredient cordycepin (FIGURE 7.), although other bioactive compounds may also be involved. Cordycepin is a derivative of the nucleoside adenosine that improves tissue blood circulation by increasing blood flow to organs such as the heart and lungs.

Other medicinal properties of the fungus include antitumor, anti-inflammatory, and antioxidant activities. For example, when metabolised to cordycepin triphosphate, the active compound interferes with cellular chemical signals responsible for enhancing carcinogenesis. This has resulted in the use of caterpillar fungus extract in cancer treatment. It is also claimed that the fungus is capable of promoting increased physical stamina and sharpened mental acuity. In addition to this multitude of alleged medicinal properties the caterpillar fungus is also purported to relieve erectile dysfunction, a claim which is probably linked to its role as an aphrodisiac and resulting in the alias ‘Himalayan Viagra’.

It is, therefore, not surprising that the fungus attracts very high prices in a market that has been subject to a good deal of notoriety in recent years. Although there have been attempts to outlaw the product in some countries, it remains legal almost everywhere. For example the caterpillar fungus trade was legalized in Nepal in 2001, which led to a sharp rise in demand. However, the black market for the fungus has elements of notoriety and danger not unlike some aspects of illegal drug trades. Driven by this poorly regulated and often dangerous market, the current price of the caterpillar fungus has been known to reach and sometimes exceed £20,000/lb. There is little wonder, then, that another nickname for the fungus is ‘Himalayan Gold’.

Climate change also seems to have influenced the growth of the cold-weather-loving caterpillar fungus. The caterpillar fungus is reported to prefer temperatures below 69 degrees Fahrenheit, and stops growing at 77 degrees Fahrenheit. Warmer temperatures have, therefore, shortened the growing season of the fungus, resulting in a decrease in its availability. This, in turn, has further inflated the price of the fungus on the black market.

Although the recreational and medicinal aspects of the caterpillar fungus continue to drive human markets that are both sanctioned and unsanctioned, there remains a good deal of interest in the role that cordycepin plays in the natural environment. At the molecular level cordycepin substitutes for the nucleoside adenosine inhibiting the role of adenosine in the synthesis of RNA

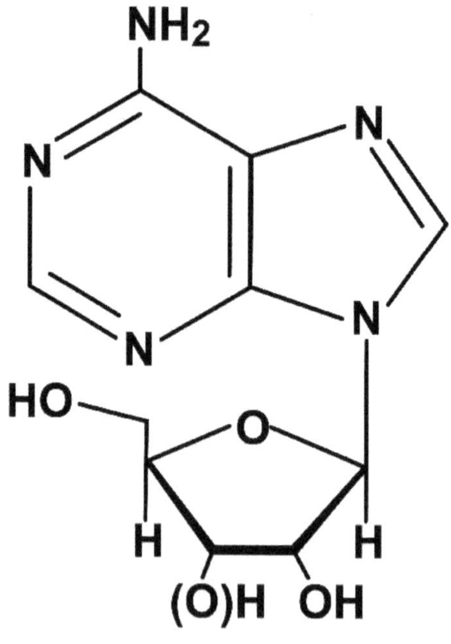

Cordycepin (3' deoxyadenosine).

In addition to moth caterpillars there are approximately 200 species of *Ophiocordyceps* fungus that are capable of invading a wide range of larval and adult insects from as many as 10 different orders. Spiders have also been victimised by the fungus. The most recently described example of a spider-eating fungus is *Purpuriocillium atlanticum*, (Family Ophiocorycipitaceae), which was first discovered in 2025 from the Amazon rain forest in Brazil as a parasite of the Brazilian Trapdoor Spider, a tarantula species noted for its tree bark habitat. Symptoms of infection range from what has been called "zombification" caused by fungal invasion of the neural system, to a more direct robotic takeover and manipulation of the musculature of the infected insect or spider. The reason for this incapacitation of the host remains somewhat uncertain, although it is reasonable to assume that a weakening of a potential host will increase the chances of multiple infections. Some infected insects display a zombie-like behavioural pattern that actually brings the prey into close proximity with the preferred habitat of the fungus, thus improving the chances of fungal infection.

An example of this is the parasitism and subsequent consumption of carpenter ants by the fungus *Ophiocordyceps unilateralis*, which is appropriately named the zombie-ant fungus, which was originally discovered by the British naturalist Alfred Wallace in 1859. Carpenter ants are members of the genus *Camponotus* a tropical ant genus native to tropical countries where they typically inhabit arboreal nests. Following their encounter with ants on the forest floor, spores of *O.unilateralis*, employ enzymes to penetrate the exoskeleton of the ant. Bioactive compounds released by the fungus in the ants' brain area result in changes in behavioural activity whereby these 'zombified' insects abandon their tree-borne nests and foraging trails and, instead, drop to the forest floor, thereby encountering a habitat with optimal temperature and humidity for fungal growth. There, they attach themselves with their mandibles to a vein on the underside of a leaf, where they remain until their death 4–10 days later. The fungus then enters the reproductive phase wherein fruiting bodies erupt from the ant's head and other parts of the corpse (FIGURE 8.), rupturing to release the fungus's spores.

The zombie-ant fungus *O .unilateralis* may be devastating to ant colonies, although *O.unilateralis* is itself also susceptible to fungal and bacterial infection, which may limit its impact on ant populations. As a protection against bacterial infestation of the fungus during reproduction *O.unilateralis* has been shown to carry antibacterial compounds that provide some defence against bacterial pathogens. As is often the case with symbiotic and parasitic organisms, the protective properties of the invasive fungus can also translate into bacterial protection for host organisms. In the case of *O. unilateralis,* the protective bioactive agents produced by the fungus have been identified as polyketones. Bioactive polyketones have now also been isolated from various fungi including *Alternaria, Aspergillus* and *Panus* species. The relatively recent identity of fungal secondary metabolites with therapeutic properties has generated a line of enquiry into the potential use of these, and similar groups of fungus as antibiotics, anticancer, anti-infective and immunotherapeutic agents (Khajuria et al. 2024).

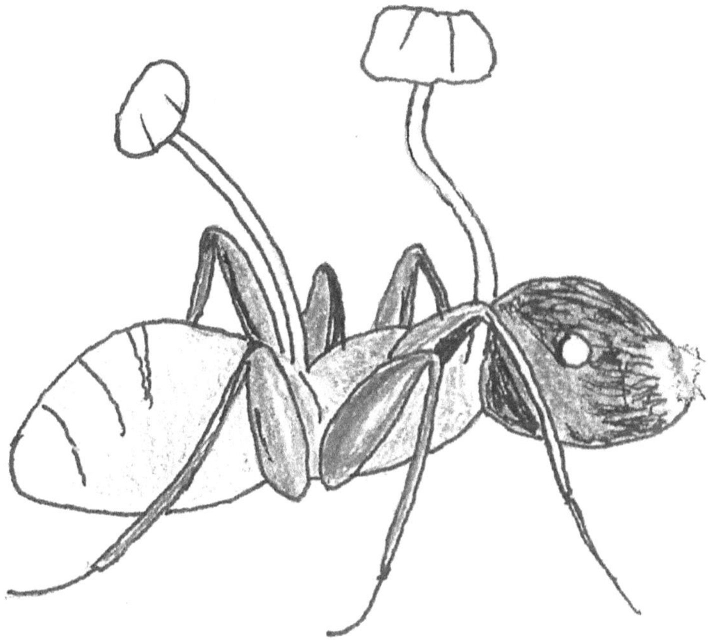

FIGURE 8. Ophiocordyceps unilateralis,

Another example of biological warfare involves the fungal pathogen *Cryptococcus neoformans*, which is a yeast belonging to the Basidiomycetes class. This parasitic fungus can live in both plants and animals including humans. In humans it can cause several different diseases collectively known as cryptococcosis whose symptoms vary according to the site of infection. Lung disease caused by *C. neoformans* is similar in many respects to pneumonia, whereas brain infection, which also originates from yeast inhalation, results in a disorder similar to meningitis. The *Cryptococcus* strain causing this specific disease is appropriately called

Cryptococcus meningitis, which is responsible for approximately 220,000 new cases and 180,000 deaths annually. Cryptococcal lung disease is generally attributed to the inhalation of fungal spores from *C. neoformans* and is exacerbated by immune deficiency, particularly that associated with HIV.

C. neoformans was first described in 1895, when it was originally split into two different variants, *C. neoformans grubi and C. neoformans gatti.* Since then, the fungus has undergone numerous taxonomic changes including its division into as many as seven distinct species! Part of this taxonomic confusion results from complexities in the identification and differentiation between the sexual and asexual reproductive stages of the fungus. The sexual reproductive form i.e. the fruiting body is called the telemorph, whereas the asexual reproductive form, which has mould-like or yeast-like appearance and is known as the anamorph. For species that only reproduce either sexually or asexually at any one time the appearance of only one of these forms is a clear source of taxonomic confusion which continues to this day. Examples of this complexity are inherent in some historic accounts of fungal taxonomy that have led to confusion over the identities to filamentous (telemorph) and yeast-like forms of the *Cryptococcus* genus. In the late 19th century *Cryptococcus* was originally wrongly identified as a member of the *Saccharomyces* (Brewer's Yeast) genus. This misidentification persisted until 1901 when they were reclassified into the *Cryptococcus* genus by the French mycologist Jean-Paul Vuillemin, who demonstrated that, while *Saccharomyces* could initiate the fermentation of sugar, *Cryptococcus* had no such fermentation capability. Further confusion arose over the initial identification of the teleomorph (sexual) and anamorph (asexual) forms of the fungus as two separate distinct species *Filobasidiella neoformans* and

Cryptococcus neoformans. However, the International Code of Nomenclature for algae, fungi and plant currently recognises *Filobasidiella* and *Cryptococcus* as synonymous.

Cryptococcal infection usually results from the inhalation of the yeast-like form of the fungus, which is commonly distributed in dried bird droppings. Lung infection in humans is usually countered by immune cells known as macrophages which act as a defence mechanism by engulfing foreign bodies such as *Cryptococcus* spores, a process known as phagocytosis. Some *C. neoformans* can survive inside alveolar (lung) macrophages, causing a degree of latency in the development of the disease, and some resistance to antifungal agents. *C. neoformans* may also combat defensive macrophage activity by up-regulating the expression of genes involved in the oxidative stress response. This process promotes genomic repair in fungi that have been subjected to the oxidative, DNA-damaging environment of the host macrophage. This activity enhances the genomic repair capability of the fungus, strengthening its defence against the host's alveolar macrophage and augments the virulence of the fungal pathology.

Other offensive measures employed by the fungus involve a high degree of genomic plasticity, including the production of massive numbers (i.e. many billions) of haploid basidiophores (reproductive spores) capable of overwhelming the defensive measures by infected organisms This process is referred to as monokaryotic fruiting.

Tinea corporis (ringworm) is a fungal infection responsible for a variety of skin disorders including Athlete's foot which is sometimes referred to as foot ringworm. *Trichophyton rubrum* accounts for 80-90% pathogenic strains. Athlete's foot is a dermatophyte skin disorder generally referred to as *Tinea pedis*, that is caused by a number of different funguses including species of Trichophyton, Epidermophyton genus

Microsporum. Division Ascomycota. Class, Eurotiomycetes. Although originating in South East Asia, and part of Africa and Australia, they are now considered to affect 3% of the world's population. Although they normally affect keratinised tissues such as skin, hair or nails, but may also infect other tissues such as mouth, lungs, throat and urinary tract.

A Fatal Fungal Mould.

Scopulariopsis brevicaulis (FIGURE 9.) is a widespread filamentous mould, also belonging to the division Ascomycota that exists in the natural environment as a soil saprophyte[15], where it acts as an agent of biodeterioration. However, it can also act as an opportunistic pathogen on animals and humans. In the human world it mainly occurs on damp walls, textiles and furniture products as well as a variety of food items. It can be isolated from cereals, legumes, nuts and apples as well as meat and dairy products.

S.brevicaulis was first described in 1882 and initially identified by the Italian botanist Pier Saccardo as *Penicillium brevicaule*, but was renamed *Scopulariopsis brevicaulis* by French mycologist Georges Bainier in 1907. Human pathologies associated with *Scopulariopsis brevicaulis* include infection of finger- and toenails (onychomycosis) leading to deterioration of nail tissue and inflammation of surrounding skin. *S.brevicaulis* is one of several genera of fungi that can cause invasive skin infections. Others include *Acrimonium, Fusarium* and *Paecilomyces.* More serious pathologies, and even death may occur following *S.brevicaulis* infection of internal tissues and organs.

---

[15] A Saprophyte is an organism that feeds on dead or dying organic matter.

A remarkable pathological effect of *S brevicaulis* was uncovered by British organic chemist Frederick Challenger in 1945. Challenger found that, while feeding on starch wallpaper paste, the fungus caused the transformation of green copper arsenate dyes that were used to colour the wallpaper into trimethylarsine oxide. The oxide is then reduced to the highly toxic trimethylarsine gas which was responsible for some human fatalities.

FIGURE 9. ***Scopulariopsis brevicaulis.*** A filamentous mould.

In recent years the nomenclature of this fungus has undergone a change from *Scopulariopsis brevicaulis* to M*icroascus brevicaulis*, yet another example of the complexities involved in fungal taxonomy. The change follows the recognition that the name *Scopulariopsis* describes the asexual form of the fungus, i.e. the anamorph, whereas the sexual form of the fungus (teleomorph) is known as *Microascus*. Because the teleomorph is the primary bioactive form of the fungus responsible for onychomycosis, the generic name *Microascus* is now preferred.

The Petrol Cap Fungus.

Following Ink Caps, Green Caps and fungal caps of various shapes and sizes, you might assume that the 'Petrol Cap Fungus' is yet another mushroom cap bearing a resemblance to a car's fuel cap. However, in this case we really are talking about the petrol cap of a car, and the mould that grows in and around it. The mould in question is *Cladosporium resinae*, which is a hydrocarbon-utilising filamentous fungus capable of feeding on hydrocarbons such as petroleum. As such, it is a remarkable example of a fungus with the ability to extract nutriment from a source (petroleum) that most organisms would find impossible to metabolise and also highly toxic. Its ability to feed on refined hydrocarbons has been recognised by the common names 'Kerosene Fungus' and 'Diesel Bug'. In addition to a carbon source, the only other requirement of the fungus is moisture, which is a commodity that is rarely far away from a vehicular petrol cap (FIGURE 10.).

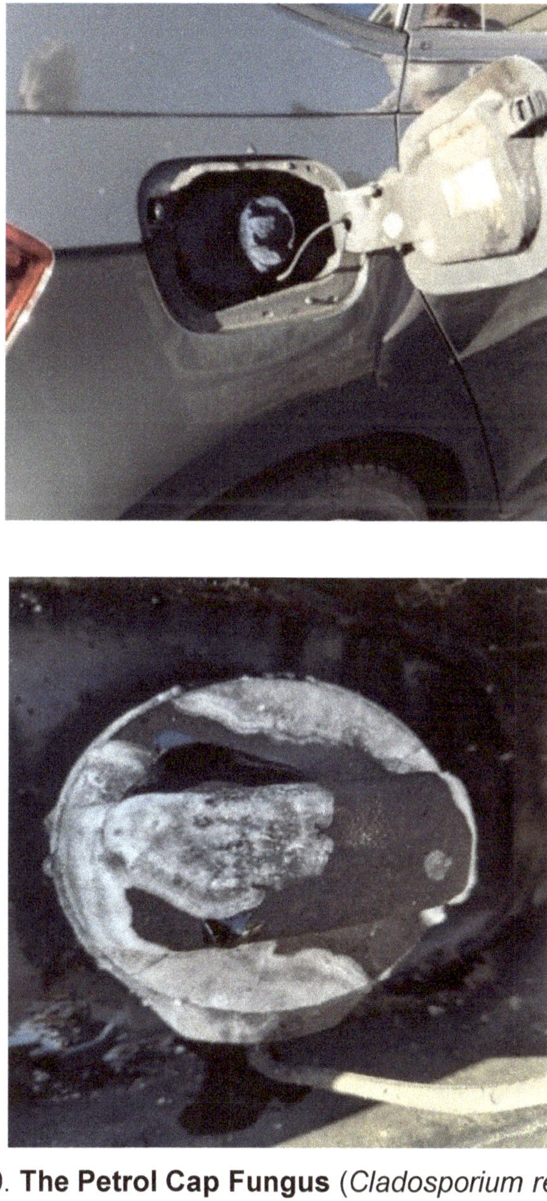

FIGURE 10. **The Petrol Cap Fungus** (*Cladosporium resinae*).

In addition to the highly specialised species *C. resinae* the genus *Cladosporium* has been expanded over the last 70 years to accommodate over 750 fungal species, many of which are plant-parasitic or saprotrophic on living or dead plant tissue. Some even parasitise other parasitic forms such as rust fungi. This large genus now has several synonyms including *Acrosporella*, *Heterosporium*, *Myxocladium*, *Sporocladium*, *Polyrhizium* and *Hormodendrum*. Several of these have become part of the human fungal biome, although they are rarely responsible for serious pathogenic conditions in humans. Nevertheless, spores from *Cladosporium* species have been reported to cause nasal congestion, coughing and some skin infections.

Dead Man's Fingers.

The genus *Xylaria* is a group of about 100 species belonging to the phylum of a fungal group known as Ascomycetes (division Mycota). They are common in woodlands in Europe, North America and other continents where they are normally found growing on or near the decaying roots of trees and other plants. They take a variety of forms. One of the most macabre species is *Xylaria polymorpha*, known as Dead Man's Fingers (FIGURE 11.). This eerie name derives from the imagined appearance of the fungus as the fingers of a buried dead man reaching up from below the earth. This creepy image derives from the finger-like fruiting bodies that protrude from the soil. These finger-shaped structures 1-4" long and 0.5-1" in diameter are usually dark grey or black often with pink or white tips. In countries such as Lithuania, they have even adopted a persona in the form of the Baltic god of the dead, Velnias.

FIGURE 11. **Xylaria polymorpha.** Dead Man's Fingers.

Another species of *Xylaria* that illustrates the extraordinary morphological diversity of this genus is *X. hypoxylon* (FIGURE 12) which is a species of fungus in the family Xylariaceae that was first described by Carl Linnaeus in 1745. This species is considered to have has several different common_names including 'the candlestick fungus', 'the candlesnuff fungus' and 'the stag's horn fungus'. Yet another example of the flexibility, some might say uncertainty, attached to its taxonomy. The fungus grows in clusters on decaying hardwood in the form of branched fruiting bodies 1-3" tall and 0.5" in diameter It is considered by some mycologists to have very slight bioluminescence, and as such is considered the only fungus outside of the Agaricales order to demonstrate this characteristic.

Xylaria hypoxylon is considered by some to have some anti-viral and antitumour properties, although it is generally regarded as inedible due to its small size and tough consistency.

FIGURE 12. **Xylaria hypoxylon. C**ommon names 'Candlestick or 'Stag's Horn Fungus'

Hazel Gloves Fungus.

The Hazel Gloves Fungus *Hypocreopsis rhododenri* (FIGURE 13.) is an Ascomyte fungus (phylum Ascomycota; class Sordariomycetes; order Hypocreales; family Hypocreaceae) that was first identified in 1922 growing on a rhododendron bush by American mycologist Roland Thaxter.

This discovery prompted the species name *rhododenri*. However. the fungus does not appear to grow on the rhododendrons of Europe where the most common habitats are temperate rainforests with oak trees as a favoured host. The name 'Hazel' in the fungus' common name appears to relate to its colour i.e. 'hazel brown' rather than the identity of its host tree, i.e. oaks rather than hazels. Less macabre than 'Dead Man's Fingers', the finger-like shape of the fruiting bodies is incorporated the more benign common name of the fungus; 'Hazel Gloves'. The 'glove' analogy is further supported by the dimensions of the fungal fingers, 0.75" in diameter X 4" long.

An interesting feature of the Hazel Gloves Fungus relates to its parasitic habitat. Although it appears as a direct parasite of its tree host, the Hazel Gloves fungus has the unusual distinction of being a parasitic fungus of a parasitic fungus. The true host of the Hazel Gloves Fungus is, in fact, the Glue Crust Fungus *Hymemochaete corrugata*. This intermediate parasite of the host tree is often hidden beneath the hazel gloves.

FIGURE 13. ***Hypocreopsis rhododendri.*** The Hazel Gloves Fungus.

Hoof Fungus.

The 'Hoof Fungus' (*Fomes fomentarius*) (FIGURE 14.), also known as the 'Tinder Fungus' is a parasitic bracket fungus[16], named for its resemblance to a horse's hoof, that is commonly found clinging to the bark of trees from many parts of the globe including Africa, Asia, Europe and North America. The caps of the fungus may range in diameter from 3 cm to more than 20 cm, the size of a dinner plate. Colours range from various shades of grey and brown.

The Hoof Fungus has the reputation of being one of the earliest fungi to be harvested by humans and has been traced to Mesolithic sites that are at least 8000 years old. The most notable of these discoveries were of two species of the fungi found in the possession the Neolithic traveller, "Ötzi the Iceman[17]" whose remarkably well-preserved 5,500-year-old body was retrieved from ice in the Austrian Ötztal Alps in 1991. Hence the name Ötzi. Also in Ötzi's possession was the Birch Polypore[18] (*Fomitopsis betulina*) another fungus with a wide variety of uses. Neither the Hoof Fungus nor the Birch Polypore are regarded as edible, although both have numerous properties, medicinal and otherwise, that would have appealed to an itinerant Neolithic traveller operating in difficult terrain and a cold climate. For example, since the time

---

[16] A Bracket Fungus is one that grows on trees.

[17] "Ötzi The Iceman", is the name given to the frozen mummified remains of a man who lived between 3350 and 3105 BC. Ötzi's remains were discovered 1991, in the Ötztal Alps close to the Austrian-Italian border (hence the name "Ötzi". The body is Europe's oldest known natural human mummy, and a unique specimen of a Chalcolithic (Copper Age) European.

[18] A Polypore is a general term referring to bracket or shelf fungi that typically grow on trees often with pathological effects. Many of these fungi have been ascribed to the Order Polyporales in the Division Basidiomycota.

of 5th century physician and philosopher Hippocrates the Birch Polypore has possessed a reputation as an antiseptic and disinfectant, and the antibacterial and anti-inflammatory properties of the Hoof Fungus have also been part of folklore for many millennia. Another remedial property of the Hoof Fungus was its ability to cauterise wounds by restricting blood flow a property made it useful as a dressing capable of staunching blood flow in case of injury. The medicinal use of this fungus as a wound dressing was recognised by the 5th century physician Hippocrates.

Non-medical, physical properties of the Hoof Fungus that would have appealed to "Ötzi the Iceman" relate to the physical strength and abrasiveness of parts of the fungus itself. It has been speculated that, hardened mature fungal plates cut from the fungus may have been employed as strops capable of sharpening weapons such as those produced in the Bronze Age and probably wielded by the Iceman.

Another practical use for the Hoof Fungus related to the cold alpine country that was home to the Iceman. Whether or not the weapons, including axes, employed by this Neolithic wanderer would have been used for hunting or for self-defence, another high priority would have been keeping warm.

Although we can speculate whether or not the Iceman would have used his newly-sharpened hardware to cut and chop firewood, an easier and proven source of fuel was at hand in the form of Hoof Funguses and other similar bracket fungi. There is evidence going back as far as 8000 years that large plate fungi were used as Bronze Age kindling and tinder. Hence the name Tinder Fungus.

It is, therefore, not difficult to imagine the Iceman using mushrooms as fuel for generating and sustaining a fire, not only as a straightforward heat source, but also for cooking. It is also not unreasonable to imagine the use of hot immersive fluid in the creation of dressings and extracts for a broad range of treatments over many millennia.

FIGURE 14. **Hoof Fungus.** *Fomes fomentarius*. Photograph by L. Bosco.

Fungal Fashion.

Whether or not the Iceman used the Hoof Fungus for protective clothing there is clear historical evidence from Eastern Europe and North America going back several hundred years of its crafting and manipulation to form a variety of items such as bags, pouches and clothing such as shoes and belts. Spongy tissue from Hoof Fungal plates is prepared for this purpose using a boiling and stretching process, to produce a soft leathery material called Amadou that is remarkable for its strength and durability as well as its lightweight nature and biodegradability. It also possesses some antimicrobial qualities related to the defence of the Hoof Fungus against potential pests and consumers in the natural environment. Other fungal tissue that has been successfully made into clothing include the mycelial network which can be trained using a pattern to form a flexible seamless mesh requiring little or no sewing to create an article of clothing. Some fungi can be used to soften clothing fabric for both aesthetic and manufacturing reasons. An example is the use of the cellulase enzyme produced by the fungus *Trichoderma reesei*. The cellulase serves to degrade denim fabric sufficiently to enhance what might be described as a "used look". In the laundry department lipases and proteases produced by the fungi Rhizopus and Aspergillus are common ingredients in most (perhaps 90%) household detergents.

Trichoderma viride is a fungus that is used as a seed dressing and soil treatment to suppress several diseases caused by fungal pathogens such as *Pythium* and *Armillaria* and for control of soil-borne plant pathogen diseases e.g. *Macrophomina phaseolina* & *Fusarium* sprcies

Mycal Medicines.

Many of the medicinal properties attributed to the Birch Polydore are now known to be shared by other similar fungi such as *Ganoderma,* which is a large tropical genus of bracket fungi native to Eastern Asia, The *Ganoderma* genus belongs to the family Ganodermataceae, which comprises about 80 species of mushroom. *Ganoderma lucidum* (FIGURE 15) was first described taxonomically by the Finnish mycologist, Peter Karsten in 1881. It is a large brownish red bracket fungus, often more than 12" in diameter, largely found growing from the base of trees in warm tropical environments.

FIGURE 15. **Ganoderma** is a bracket (shelf) fungus

Polysaccharides and terpenes from *Ganoderma* species have been linked to a broad range of health benefits including antimicrobial, antioxidant and antitumour activity. These claims follow a long medical history of the mushroom *Ganoderma lucidium* as a remedy for many ailments, emanating primarily from China, Japan, Vietnam and India. Demand for the mushroom capitalises on an ancient Chinese recipe going back at least 2000 years which designated this fungus the Reishi Mushroom 'The Food of Immortality'. Another name for the mushroom, 'The Food of the Emperor', was clearly a comment on the fact that only the rich could afford this delicacy!

It has a rich cultural history. Carvings of this fan-shaped bracket fungus, known in China as *Lingzhi* or 'the Tree of Life', adorn the walls and furniture in Beijing's Forbidden City as well as other works of art in several Asian countries. Its divine properties have even resulted in the mushroom's head being sculpted into the Chinese Emperor's wooden sceptre.

These historic imperial legacies coupled with the ostensibly wide-ranging curative properties of Reishi mushrooms triggered the marketing of Ganoderma coffee as a health drink by Leow Soon Seng in Malaya in the late 1990s. Although its efficacy remains a source of scepticism in some reviews, Ganoderma coffee is now the focus of a multi-billion-dollar trade in Eastern Asia and elsewhere.

The Reishi mushroom (FIGURES 16 and 17) has been credited with a very broad range of medicinal properties including enhancement of immune response, anticancer, antidiabetic, antioxidant, analgesic, anti-inflammatory, antimicrobial as well as estrogenic activity.

As well as its purported medicinal properties the Reishi mushroom has also gained a reputation as a trinket with sexual connotations; not so much an aphrodisiac, more of a gift given by men to women in whom they had a sexual interest.

To satisfy popular demand for this medicinal mushroom there is a market for the sale of immature assemblages of the fungus before the fruiting bodies of the mushroom colony have fully developed (FIGURE 16). The mature, edible form of the mushroom is a glossy brown (FIGURE 17).

FIGURE 16. Early Growth Phase of the **Reishi** mushroom.

FIGURE 17. **Reishi Mushroom.** The mature form of the mushroom

The Turkey Tail Mushroom.

The Turkey Tail Mushroom, *Coriolus versicolor* (FIGURE 18.), also known as *Trametes versicolor* and *Polyporus versicolor* is a saprophytic bracket fungus that grows on the dead and dying trunks and branches of hardwood trees such as beech and oak throughout the world. It was originally described by Carl Linnaeus in 1753 as *Boletus versicolor*, although its name was changed to *Trametes versicolor* by Czech mycologist Albert Pilát in 1936. The Turkey Tail mushroom is found in a range of climates from tropical to temperate, and its irregular lobed structure consisting of overlapping tiers coupled with concentric zones of colour makes it one of the most easily recognised and popular mushrooms in the world. However, its popularity stems mainly from its aesthetic characteristics rather than its culinary qualities. Although Turkey Tail mushrooms are reputed to have a milder, less bitter flavour than Reishi mushrooms and are non-poisonous, they are usually described as 'inedible' due to their tough leathery consistency. Nevertheless, they have a long remedial and pharmaceutical history going back more than 2000 years. The use of these mushrooms as a treatment for numerous medical disorders is recorded in Chinese medical records as old as 200 BCE. Knowledge of these ancient remedies comes to us from the *Bencao gangmu*, better known in English as the *Compendium of Materia Medica* or *Great Pharmacopoeia,* an encyclopaedia of Chinese herbology written by Li Shizhen and finally published in 1596. The descriptor 'Turkey Tail' applied to the mushroom is a much more recent name than any found in ancient Chinese texts and probably derives from the United States within the last few hundred years. The term clearly refers to the similarity between the radial patterns of the mushroom cap and the splayed tail feathers of a turkey.

The question arises "how can the medicinal use of Turkey Tail mushrooms be reconciled with their description as 'inedible' "? The answer is also found in the *Bencao gangmu*, where the medicinal properties of the mushroom are attributed to 'a tea that is beneficial to the spirit and vital energy and capable of strengthening bone and tissue'. In other words, no attempt needs to be made to digest the mushroom cap when its remedial powers can be transferred to an infusion of the fungal tissue in hot water.

Since the 1960s the medicinal uses of the Turkey Tail fungus and its extracts have been expanded to take advantage of its purported antioxidant properties and remedial power over several immunological disorders. In Japan and China the Turkey Tail mushroom extract has been credited with the remission of various types of metastatic cancer. The active ingredients of the fungus have been identified as the β-d-glucan polysaccharide-K (PSK) and Polysaccharide Peptide (PSP). Since 1983, PSK and PSP have been approved in China as therapies for and treatment of lung- and gastrointestinal cancer.

FIGURE 18. **Turkey Tail Fungus** *Coriolus versicolor*.

Chicken of the Woods.

The popularity of several species of bracket mushrooms owes more to their flavour than their medicinal properties. A well-known example is the bracket mushroom *Laetiporus sulphureus* (FIGURE 19.). Although this mushroom has demonstrated significant anticancer activity and apparently possesses some antibacterial, antimicrobial and pesticide qualities, its primary human attraction is signified by its common name “Chicken of the Woods”. The name derives from the chicken-like flavour of the fungus, which is regarded as a culinary delicacy in the areas of North America, Africa and Europe where it grows as a weak parasite on several different trees notably beech, chestnut and oak trees. *Laetiporus conifericola* is a closely related species found on coniferous trees.

Laetiporus sulphureus was first named *Boletus sulphureus* by French mycologist Pierre Bulliard in 1789. In common with many other fungi it has had many synonyms and taxonomic updates, and was finally given its current taxonomic name *Laetiporus sulphureus* in 1920 by American mycologist William Murril. 'Laetiporus' is translated as meaning "with bright pores" and *sulphureus* means "the colour of sulphur. Other names for the fungus include sulphur polypore and sulphur shelf. The name ‘sulphur’ refers to the bright yellow colour of the fruit bodies of the fungus as they appear in its younger and most edible stage. Older specimens tend to be less digestible and are paler in colour with the centres of the “shelves” adopting more of a pinkish orange tinge (FIGURE 19). All “Chicken of the Woods” fungi require cooking. The fungus is prepared, cooked and served much like chicken, along with the condition that it should not be eaten raw or in large portions at risk of adverse gastrointestinal side-effects. It seems reasonable to assume that the substitution of the fungus for chicken is particularly attractive to followers of a vegan diet!

So far, we have seen how fungi have invaded numerous plants resulting in both beneficial and adverse effects on those plants. However, in many cases the good and bad consequences of fungal infection are not mutually exclusive. Fungi are commonly regarded as parasitic and damaging to tree bark and the roots of trees and other plants. From the human perspective much of the focus on fungi has been related to their adverse effects on plants that are of commercial value. These include agricultural crops and plants grown in park land and for recreational purposes. However, most human interest on fungi has focussed on their attraction as culinary delicacies or their medicinal properties.

FIGURE 19. **Chicken of the Woods** *Laetiporus sulphureus*.

Edible Parasol.

It would be difficult to find a more popular fungus than the parasol mushroom *Macrolepiota procera*. This large saprotroph appears either as an individual or as a clump of mushrooms in temperate grasslands in Australia, East Africa, East Asia, North America and Europe. For the most part it lacks an intimate mycelial relationship with plant roots but may still act as a decomposer of tree-root bark in woodland areas.

The mushroom was initially given the name *Agaricus procerus* by the Italian Giovanni Scopoli in 1772 but was reclassified by Samuel Gray as *Lepiota procera* in 1821. In 1948 it was given a new generic name *Macrolepiota* by German mycologist Rolf Singer to accommodate its unusually large size. Standing up to 9" tall the mushroom has an elegantly patterned cap which expands from a spherical shape into a round flat 'parasol' which may reach 12" in diameter (FIGURE 20). The species name is derived from the Greek 'Lepiotos', meaning scaly, in reference to its elegant scaly patterning that adorns the flat cap at maturity (FIGURE 20.). The scales are characteristically brown against a whit or pale pink background. Another characteristic feature of the mushroom is a double-edged ring appearing high on the stem. The stem itself has an elegant sculptured texture.

FIGURE 20. **The Parasol Fungus** *Macrolepiota procera*

Apart from its aesthetic appeal the mushroom is regarded as a culinary delicacy. Its nutty flavour coupled with its large size place it in high demand as a gastronomic treat. In addition to its savoury appeal, it is also regarded as a useful source of important trace metals such as zinc and copper However, despite its savoury attraction, a note of caution attached to the parasol mushroom is its similarity to some related species such as *Chlorophyllum racodes* that can cause stomach upset if eaten raw or undercooked.

Black (or White?) Rot.

The Chaga fungus *Inonotus obliquus*, (FIGURE 21.) is a member of the fungal family Hymenochaetaceae (order Hymenochaetales; class Agaricomycetes). The family Hymenochaetaceae is a taxonomic division comprising about 34 genera and 670 species. The taxonomy of many of these fungi has been defined primarily on the basis of molecular phylogenetics over the last 20 years or so, rather than the physical characteristics the fungi themselves. Indeed, it would be difficult to equate the Chaga fungus with other fungi based on its unique physical characteristics. The fungus usually appears as a black mound, often 12" or more in diameter on the damaged bark of alder, beech and (mainly) birch trees (FIGURE 19). This black mass, known as the conk, is actually the mycelium of the fungus which invades the bark of trees at areas where physical damage has occurred. Following invasion the fungal fruiting bodies are hidden within the tree, which means that the conk is only visible part of the fungus. This mass has often been likened to burnt charcoal because of its increasing pigmentation by melanin as the conk grows. The fungus contributes to the deterioration of the bark on which it feeds and may cause serious damage to the tree in the form of white rot.

Although the chaga fungus has been shown to damage and even kill some tree species in cold climates their black masses, or conks, have nevertheless been sought after for their medicinal properties in powder or liquid form. Alleged remedial properties include antioxidant, anti-cancer and immune enhancement, although reliable clinical data are illusive. However therapeutic properties are claimed for *Inonotus obliquus* which is now commercially marketed for medicinal properties.

FIGURE 21. ***Ionotus obliquus*** ("chaga") is a tree parasite

The Sulphur Tuft

The Sulphur Tuft mushroom *Hypholoma fasciculare* (phylum Basidiomycota; class Agaricomycetes; order Agaricales; family Strophariaceae) (FIGURE 22) is generally found growing on the dead wood of deciduous in Northern Europe, Asia Minor and parts of North America. To a lesser extent they also appear on some conifers.

The Sulphur Tuft was first identified in 1778 by British botanist and mycologist William Hudson who gave it the name *Agaricus fascicularis.* However, it was reclassified by German mycologist Paul Kummer as *Hypholoma fasciculare* in 1871. The species name *fascicularis* is derived from the Latin word 'fasce' meaning bundles, with perhaps a hint of 'strength in numbers' about the definition. This description refers to the frequently large clusters that are a feature of this mushroom's common woodland appearance throughout their range. Another noticable characteristic is their bright yellow colour which is responsible for the name 'sulphur'.

FIGURE 22.. **The Sulphur Tuft mushroom** *Hypholoma fasciculare.*

Despite its attactive appearance the Sulphur Tuft mushroom has a very bitter taste with several adverse health effects following accidental consumption or mistaken identity. The toxicity of these mushrooms has been attributed, at least partially, to the toxic steroids known as fasciculols.

Fasciculols are multi-cyclic steroids having the general formula shown in FIGURE 23 but with the substituents at positions R1 and R2 either hydrogen or the sub-group shown as FIGURE 24.

Fasciculols have been shown to be calmodulin inhibitors. Calmodulins are calcium-binding messenger proteins that mediate numerous crucial metabolic processes such as apoptosis, kidney function (proteinuria), inflammation, smooth muscle contraction, intracellular movement, and immune response. In humans, symptoms may be delayed for 5–10 hours after consumption, after which time there may be diarrhoea, nausea, vomiting and convulsions. Paralysis and impaired vision have also been recorded.

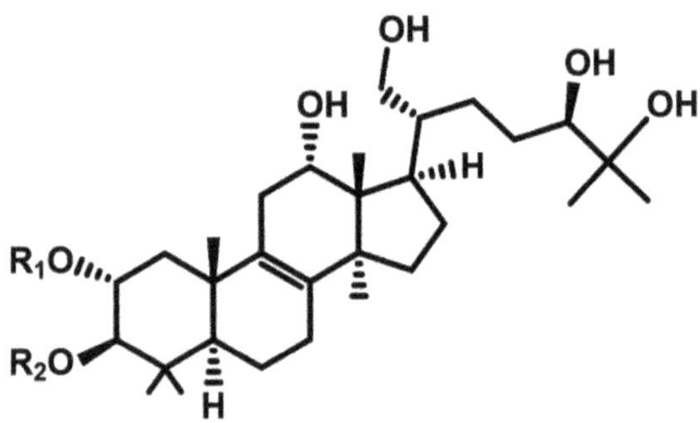

FIGURE 23. General formula for steroids known as fasciculols, which are toxic constituents of the Sulphur Tuft fungus. Sub-groups at R1 and R2 positions may be either hydrogen atoms (H) or the sub-group shown below paired with H (FIGURE 24 below). Courtesy Creative Commons Corp. Mountain View, CA, U.S.A.

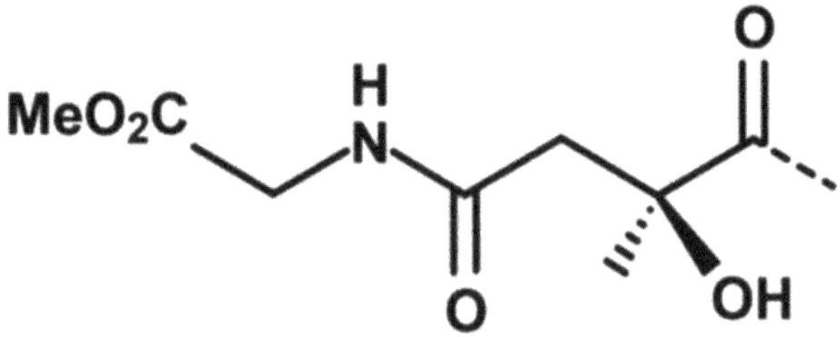

FIGURE 24. Molecular Sub-Group common to toxic steroids (Fasciculols E and F) at R2 and R1 positions respectively (see above). Fasciculol C has H atoms at both R1 and R2 positions. Courtesy Creative Commons Corp. Mountain View, CA, U.S.A.

A Delicious Redhead?

A member of the Agaricomycetes class of the fungal Divison Basidiomycota with generally positive culinary reviews is *Lacterius deliciosus*, the Red Pine Mushroom (FIGURE 25.) otherwise known as the Orange or Saffron Milkcap. This colourful mushroom is widespread in the Pyrenees, Mediterranean, Eastern European and Asian countries where its association with pine trees and pine forest litter explains the appearance of the word ‘pine’ in its most common name. It is a frequent inhabitant of temperate forests notably in Russia where it is highly regarded and goes by the name of ‘Rhzhiki’, meaning ‘redhead’. In Russia, the mushroom is often salted as a method of preservation and is traditionally served in salted or pickled form as an appetizer along with a glass of vodka.

L. deliciosus was originally named *Agaricus deliciousus* by Carl Linneus in 1753, but today it retains the generic name *Lactarius* given in 1821 by British mycologist Samuel Gray. Although some have questioned the species name ‘*deliciosus*’ (derived from the Latin for ‘tasty’) as somewhat overenthusiastic it is still widely regarded in several countries as a delicacy rich in minerals and vitamins, as well as terpenoids with anti-tumour, antiviral and anti-inflammatory properties. In France and Russia, it has been recommended as a treatment for ailments such as tuberculosis and asthma. Frescos from ancient Herculaneum and Pompeii apparently depict the use of the Red Pine Fungus going back at least 2000 years, although it is impossible to tell whether the appeal of the fungus in ancient times was purely gastronomic, medicinal - or both. Certainly, the bright orange red colour of the mushroom caused by high levels of beta-carotene add to its appeal, as long as eaters of the mushroom have no problem with the dye finding its way into their urine following consumption!

FIGURE 25. **The Red Pine mushroom** *Lacterius deliciosus* is a large edible mushroom widespread in temperate forests whose cap changes from convex to concave when it matures.

The Horn of the Dead.

Craterellus cornucopoides (FIGURE 26) commonly known as the 'Trumpet of the Dead', 'Black Trumpet' or 'Horn of the Dead', is much less spooky than its common names suggest. Although its black funnel shape is portrayed as a herald of death in some cultures, the reality is much more benign. Far from being a portent of death that should be avoided, the mushroom is regarded as a delicacy throughout much of its range.

Despite its somewhat macabre appearance the mushroom is regarded as not unpleasant to eat and non-toxic, but also of significant calorific value and a source of fatty acids vitamins B12 and C. These properties have contributed to an alternative pseudonym for the mushroom; 'Horn of Plenty'. The name plenty is embodied in the species name *cornucopoides.* Not surprisingly the contradiction between the benevolent 'Horn of Plenty' and the macabre 'Horn of the Dead' has raised questions. One answer that has been given is that the so-called 'Horn of the Dead' simply reflects the late autumn emergence of the mushroom at the onset of winter when most vegetation is dying back'. Of course, another answer that could simply be that folklore often has a tendency towards the other-worldly and the spiritual! For this provenance we need to look no further than Amathea, the nurse of Zeus. According to 4th century BCE mythology it was Amathea's fabulous 'Horn of Plenty', supposedly derived from her goat, that was capable of producing endless amounts of food and drink.Some versions of this story have Amathea herself as the goat. Either way, this cornucopia of food and drink has found its way into the taxonomy of the mushroom' in the form of the species name *cornucopoides*.

The mushroom was first described by the Swedish botanist Carl Linnaeus in 1753 as Pieza conucopoides but was transferred to the *Craterellus* genus by mycologist Christiaan Persoon in 1825. which extends throughout East and Central Asia and Europe. It has also been reported from North America, although the American species now goes by the names *Cratellerus fallax* on the East Coast and an unspecified (*Cratellerus spp.*) elsewhere on the North American continent.

FIGURE 26. ***Craterellus cornucopoides*** This mushroom has a variety of common names including both the sinister 'Trumpet of the Dead', 'Black Trumpet' and 'Horn of the Dead', as well as the more flattering 'Horn of Plenty'. It is widespread in Europe and Eastern Asia, where it grows on damp forest floors often in association with broad-leaved deciduous trees such as beeches and oaks.

Hair Ice. A Fungal Phenomenon.

Ice, almost by definition, is unyielding, crystalline and brittle. These characteristics have formed the basis of several metaphors indicating rigidity. We are all familiar with sayings like 'as hard as ice' or 'as solid as ice' and have grown up with the familiar patterns of ice crystals that have adorned our windows in cold frosty weather. For generations these hexagonal crystalline figures have formed the basis for the crisp snowflakes featured on many Christmas cards. Throughout most of this period maxims such as 'as bendy as ice' or 'as flexible as ice' were unheard of. However, in 2015 we saw a fascinating change in our perspective on the apparent pliability of ice. This resulted from research carried out by the German scientist Diana Hofmann and her colleagues Gerhart Wagner and Christian Mätzler. The group were investigating a remarkable phenomenon that is now known by the name 'Hair Ice'. Hair Ice is uniquely found attached to decaying tree branches and other organic debris littering some European woodland floors in cold weather. As the name implies, unlike the snow-like icy covering that most of us are used to, Hair Ice takes the form of hairy wisps that bear a remarkable resemblance to white human hair. What begins as a fuzzy outgrowth from a fallen twig or branch in sub-zero temperatures (FIGURE 27a) turns into white silky strands as thin as 0.02mm. in diameter that would look 'at home' on the beard of Santa Claus (FIGURE 27b). Even on close inspection it is difficult to believe that this is ice, not hair. That is, of course, until it melts away at room temperature in just a couple of minutes! Strands of hair ice may grow as long as 20cm and look nothing like the frost that we are used to seeing on the ground and shrubs in cold weather. Its most extraordinary characteristic is the wavy, sculpted appearance of the frozen ice (FIGURE 27b).

The most remarkable aspect of this phenomenon turns out to be the involvement of a specific fungus by the name of *Exidiopsis effusa* in producing this sculpted appearance of the ice. Although the relationship between this fungus and the appearance of Hair Ice had first drawn the attention of German Scientist and climatologist Alfred Wegener in 1918, it was the work of Hoffman et al. (2015) that provided the critical information regarding the mechanism involved. This included the important observation that killing the fungus by heat-shock eliminated Hair Ice formation. It was discovered that the Hair Ice formation is caused by the presence of live fungus on rotting tree trunk tissue that has been largely stripped of bark. Under these conditions the fungus prevents the ice from crystallizing in an encrusted form. Instead, at temperatures just below freezing, water is drawn from exposed wood pores by the freezing process where it forms long delicate white strands. The phenomenon is characteristic of humid northern latitudes between 45 and 55^0N.

FIGURE 27a. **Hair Ice** begins as fuzzy outgrowths on fallen branches and leaves colonised by the fungus *Exidiopsis effusa.*

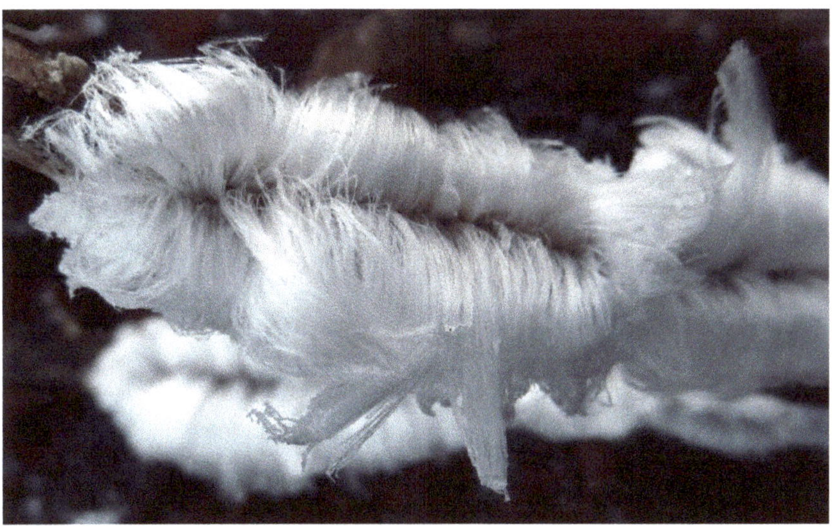

FIGURE 27b. **Hair Ice** grows in subzero temperatures as white silky strand reminiscent of Santa Claus' beard.

The genus *Exidiopsis* is a genus of about 30 species belonging to the family Auriculariaceae in the division Basidiomycota. Despite the relatively large number of species comprising the *Exidiopsis* genus, it has been reported that *Exidiopsis effusa* (FIGURE 28.) is solely responsible for the Hair Ice phenomenon. However, this remains a matter of conjecture, and it is possible that other *Exidiopsis* species may also cause the formation of Hair Ice. However, before investigating which species may or may not be responsible for Hair Ice it is necessary to comprehend the mechanism by which Hair Ice is formed. As shown by Hoffman and co-workers in their 2015 research, of critical importance is the presence of *Exidiopsis effuse* which is a saprotrophic fungus that thrives on damp decaying wood. The fungus itself consists of a gelatinous film on fallen branches, roots and leaves in many temperate forests in Europe and North America. The gelatinous nature of the fungus has given it the name jelly fungus by some observers, although accounts of its appearance vary considerably. For example, the colour of the fungal mass has been variously described as "pinkish white" (FIGURE 28.) and "light brown".

FIGURE 28. ***Exidiopsis effuse*** is a gelatinous saprophytic fungus responsible for the formation of Hair Ice on hardwood trees

There seems to be general agreement that the fungus initially occurs in patches including a mycelium that coalesce to form a continuous film. It was the observance of this fungal mycelium that first gave Alfred Wegener the clue as to the involvement of the fungus in Hair Ice formation.

The fungus contributes to the degradation of the wood by the release of cellulases and similar compounds that cause the breakdown of cellulose and lignin. This activity returns these compounds to the ecosystem where they become available as nutrients for the fungus and other life forms. In other words, *E. effusa* acts as a decomposing agent. Bearing in mind that this organic degradation activity occurs in in damp temperate climates capable of subzero temperatures in winter months it is inevitable that compounds such as cellulose and (primarily) lignin will solubilise in the surface water immediately adjacent to the woody tissues and will be present as solutes when that water freezes. It is these complex molecules, in ways that remain incompletely understood, that act as "recrystallization inhibitors". In doing so the solutes maintain a degree of segregation of ice crystals that prevents them from growing large and forming the rigid crystalline skeletons that are characteristic of frost and icicles. Instead, the crystalline structure of the ice is reduced to ultrathin filaments seemingly capable of varying degrees of flexibility.

Myca Music.

In rare cases fungi have played a part in enhancing the aesthetic and commercial value of wood products in ways that enhance their utility (and quality) for certain applications.

One of the more esoteric uses of fungal mycelia has been the treatment of specific types of wood that are used to make violins. Since the advent of the Stradivarian violins of the late 17th and early 18th centuries much effort has gone into reproducing the sound of these classic instruments at prices that are affordable. A critical aspect of this quest has been the selection of lightweight wood that combines malleability and attractive wood grain with strength and sound resonance. In an effort to obtain and maximise these qualities, Norwegian Spruce has been typically selected for the upper plate of the violin whereas sycamore is favoured for the back, sides and neck of the instrument. An important development in attaining the ideal mix of qualities for the violin's optimal musicality has stemmed from the work of the Swiss Professor of Materials Science and Technology Francis Schwartz. In 2012 Professor Schwartz demonstrated that the tonal quality of wood destined for violin manufacture could be significantly improved by pretreatment of the wood with specific fungi. The fungi selected for this procedure are the white-rot fungus

Schizophyllum commune which is used to treat the sycamore and the soft-rot fungus *Physiosporinus vitreas*, which is used to treat the Norwegian Spruce. Treatment consists of soaking plates of the wood in a growth medium containing fungal mycelia. Incubation in this medium generally takes 2-6 months and results in what is essentially a degradation of the wood. This process results in the thinning of the wood, which helps to enhance what is sometimes referred to as its elasticity, yet the wood retains its ability to conduct sound. Following the fungal immersion period the wood is treated with ethylene oxide gas to kill the fungal mycelium thereby stopping the degradation process at a point where its musicality and structural integrity of the wood is maintained. Violins made with the assistance of fungal treatment have been given the name Myco-Wood violins. While still regarded by some as a 'work-in-progress' these instruments continue to receive accolades that place them in the £20,000 plus range.

Magic Mushrooms.

The hallucinogenic qualities of several species of mushroom have been known for many hundreds of years, and it has been speculated that mushrooms have played a part in the spiritual history of mankind for several millennia. However, in part because of their mind-altering tendencies as well as their remedial qualities it has often not been easy to separate fact, folklore and fiction.

Archaeological evidence from Saharan cave paintings more than 10,000 years old depicts the superimposition of human figures and mushroom shapes. What does this mean? What is the story behind these images? Did the mushroom occupy a spiritual niche in this and other ancient cultures?

Psilocybin and Psilocin, Magic Mushroom Hallucinogens.

Psilocybin (FIGURE 29.), also known as 4-phosphoryloxy-*N*,*N*-dimethyltryptamine (4-PO-DMT) is a bicyclic alkaloid[19] compound in more than 200 mushroom species. Following ingestion, psilocybin is rapidly metabolised by phosphatase enzymes into its dephosphorylated form psilocin (FIGURE 29.) which acts as an agonist of the neurotransmitter serotonin. Psilocin acts on 5-HT_{24} serotonin receptors in the brain by displacing the normal substrate, resulting in a range of aberrant neurological responses.

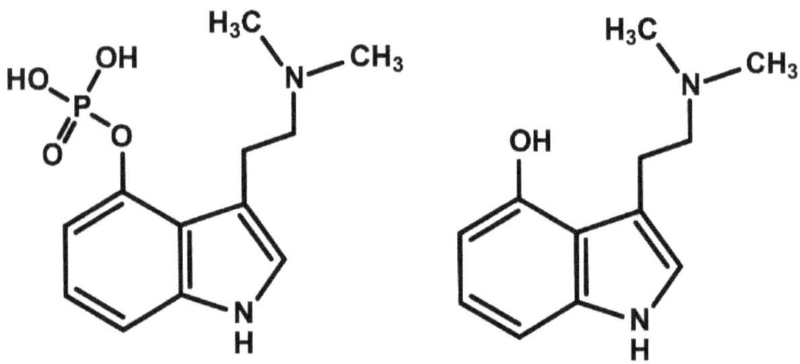

Psilocybin **Psilocin**

FIGURE 29. **Psilocybin** is a bioactive alkaloid compound in more than 200 mushroom species. Following ingestion, psilocybin is rapidly metabolised by phosphatase enzymes into its dephosphorylated form **psilocin**.

[1919] An alkaloid is an organic compound containing a nitrogen atom as an integral part of one or more carbon rings that are molecular constituents. Many alkaloids are bioactive molecules.

In many respects, Psilocin mimics the effect of other psychotropic substances such as lysergic acid diethylamide (LSD) and N,N-dimethyltryptamine (DMT). Psilocin and its molecular precursor psilocybin have lent their name(s) to this class of mushrooms, often collectively known as psilocybin mushrooms. The most potent members of the group belong to the eponymous genus *Psilocybe*, although as many as 12 other psiolocybin-containing mushroom genera are known to exist. These include *Conocybula, Galerina, Gymnopilus, Inocybe* (FIGURE 30)*, Panaeolus* and *Pluteus*. All these fungi are notorious for their hallucinogenic effects on human consumers which include a sense of euphoria and the induction of a dreamlike state involving a distorted sense of time. These symptoms are often interpreted as a magical or spiritual experience and are primarily responsible for the term 'Magic Mushrooms' that is traditionally applied to these fungi. In many cases part of the magical element associated with these and many other mushrooms stems from their frequent discovery in the form of 'Fairy Rings', i.e. circles of the fruiting bodies.

FIGURE 30. ***Inocybe aeruginascens*** is a psychoactive mushroom native to parts of Europe and North America.

Fairy Rings. Good or Bad Luck?

Although the ringlike shape of the mushroom colony (FIGURE 31) is the result of subterranean mycelial activity spreading outwards from a central location in a quest for food, folklore often tells a different story. In many countries, particularly in western Europe, these rings of mushrooms found on the forest floor are seen as the result of supernatural activity. The precise nature of this magic varies among different cultures. Some believe that anyone stepping into the fairy ring will be carried off into the fairy kingdom and will become invisible in ‘the real world’. Others simply see fairy rings as the sign of the Devil. However other communities regard fairy rings as a sign of good fortune!

FIGURE 31. A ring of *Inocybe* mushrooms. (South Wales).

In addition to psychoactive hallucinogenic responses psilocybin may also elicit more tangible tissue responses such as pupil dilation, altered (mainly increased) heart rate and blood pressure. The perceived spiritual enhancement or 'other-worldly' experience means that magic mushrooms have become a highly sought-after commodity. However, the history and nature of human use remains elusive.

Early Rock Art from Saharan desert is believed to depict that *Psilocybe hispanica* may have been used in religious rituals more than 10,000 years ago, although there is evidence of the existence of mushrooms alongside humans in all continents throughout the Stone Age. The Stone Age era has been broadly divided into three sub-divisions, the Palaeolithic, the Mesolithic and the Neolithic, although this categorisation suffers from gross differences in definition. For example the Palaeolithic period has itself been sub-, some might say sub-subdivided into an ancient prehistoric Palaeolithic era stretching back millions of years, with a Middle Palaeolithic period spanning 300,000 – 50,000 years ago. The situation is further complicated by differences in defining these ages according to the geographical source of the archaeology. The forementioned Middle Palaeolithic nomenclature subscribes to African archaeology, although European definition of these and other timelines may differ.

In view of these differences in defining epoch terminology, it is less confusing to consider the relationship between humans and fungi in terms of evolutionary events and dates for which there is general agreement. However, although we might demonstrate human and fungal coexistence there remains a good deal of uncertainty regarding the nature of that relationship.

Part of this uncertainty relates to the hallucinogenic nature of certain species of fungi. In some cases, the hallucinogenic nature of the fungi themselves may have a bearing on how they are perceived and may blur the boundary between fact and fiction particularly when viewed through the lens of time. An amusing (although by some accounts disturbing) example illustrating the pitfalls associated with this topic is now known as the 'Stoned Ape' theory.

The 'Stoned (but smart) Ape'.

The 'Stoned Ape' theory is one that has gained some traction in recent years but still remains highly controversial. This notion was initially espoused by the American botanist and occultist Terence McKenna and is described in his 1992 book *"Food of the Gods. A Radical History of Plants, Psychedelics and Human Evolution"*. The theory elaborates on the generally accepted idea that our human ancestors migrated from a forest to an open, more desert African environment approximately 2 million years ago. McKenna's theory claims that this transition was fuelled in part by the ingestion of psilocybe mushrooms that colonised the faecal droppings of cattle that inhabited these African plains. By eating these ancient ancestors of the Magic (Psilocybe) Mushroom these early hominids took advantage of the mushrooms' hallucinogenic properties by developing a generally increased awareness, heightened intelligence and creativity that became a feature of the *Homo* genus. If you feel that a degree of bias might have crept into McKenna's thinking, then you will not be surprised to learn that he was a fierce advocate and user of magic mushrooms and various other mind-bending agents including cannabis and opium. If you have ever encountered the belief that the human race and the participants' interaction with each other could be improved if everyone took a few extra mind-altering shots or concoctions on a regular basis, then you might detect a certain resonance with McKenna's theory. It is, therefore, no surprise that McKenna was a follower of the author and mystic Aldous Huxley who was a keen user of hallucinogens such as LSD, although his primary drug of choice was the psychoactive protoalkaloid mescalin, a product of the Peyote Cactus (Wright 2025). McKenna was also a disciple of the Swiss psychiatrist and philosopher Carl Jung (1875-1961), whose attempts to integrate the conscious and unconscious aspects of the psyche bear some resemblance to Mckenna's philosophy. McKenna and Jung have been strongly criticised for their beliefs, and the word 'pseudoscience' has been used by their detractors. However, whereas criticism of Jung's work was largely based on a percieved tendency towards racism in his beliefs, McKenna's critics point to a dangerous overreach in defining the role of fungal psilocybin in human evolution.

However, a more objective approach might downplay the role of magic mushrooms in driving a major evolutionary event, while recognising that their active ingredients may still have clinical importance in the treatment of certain ailments and may have a part to play in the spiritual development of different cultures. The Mexican mycologist Guzmán and collaborators (1998) published a comprehensive guide to the worldwide distribution of neuroactive fungi. *"A Worldwide Geographic Distribution of the Neurotropic Fungi.* By 2005 Guzmán had increased his estimate of the number of psilocybin-containing *Psilocybe* to 144 species.

There is little doubt that the appearance of fungi in association with their host plants long predates human existence, although an examination of the fossil record and ancient human art has shown that fungi have only been intimately associated with humans since the Stone Age era. However, the exact nature of this relationship remains uncertain.

The 'Stoned Ape' story tells us that any perceived relationship between humans and fungi remains speculative until hard evidence can be obtained. However, this is easier said than done! Even the joint appearance of mushroom and human images in Stone Age art give us no real clues regarding the actual nature of the human/fungi relationship. The simplest explanation of this human-fungal link might be that early humans and other hominids ingested mushrooms and related fungi. If this is, indeed, the case several follow-up questions arise: Was this for nutrition? Was this for ritual purposes? or were our ancestors already aware of the mushrooms' medicinal (or aphrodisiac?) properties. Of course, another answer may be "none of the above". We are already aware of the historic use of mushrooms and similar fungi as kindling. In a hunter-gatherer society that is reliant on cooking to prepare safe, edible food there are few resources more valuable than fuel for cooking fires. This is in addition to the need for our early hominid ancestors to keep warm in chilly climates, particular in places where firewood was in short supply.

Puffballs.

In addition to the obvious value of providing fuel for fires in cold weather, there are signs that, in some societies, mushroom-burning has carried a more spiritual significance, often coupled with a range of other benefits to humans. Few fungal groups better illustrate this versatility than Puffball Mushrooms.

Puffball Mushrooms are a multispecies group of mushrooms that were originally classified in the historic division Gasteromycetes, although this term has lost its taxonomic standing and puffballs are now regarded as belonging to the division Basidiomycota. The name Puffball covers three different mushroom species; *Calvatia gigantea*, (Giant Puffball), *C. lycoperdon* (Common Puffball) and *C. calbovista* (Sculptured Puffball). However, variants of these species are found in the literature and several other genera have also been given the name Puffball. These include *Abstoma reticulatum, Bovista dakotensis, B. tomentosa*, *Astraeus hygrometricus* and *Lycoperdon perlatum*

Puffball mushrooms are widespread throughout the world in temperate woodland habitats where they are often found in colonies on the forest floor. Unlike other mushrooms, the fruiting bodies of puffballs do not have gill-like spores beneath a disc- or dome-shaped cap. Instead, spores are encapsulated within a spherical cap that bursts open at the top when ripe, releasing enormous numbers of spores. Although a few puffballs have rudimentary stalks, most appear as spheres. (FIGURE 32.) The sizes of their fruiting bodies vary enormously. The caps of common puffballs are often no more than a couple of inches in diameter, similar in size to a table tennis ball, whereas giant puffballs may possess caps the size of a football, as much as a foot in diameter. Spores released by giant puffballs can number as many as five trillion!

Puffballs have gathered a large number of folk names in various cultures. Some of these make colourful reference to the digestive system and excrement. For example, the Latin name of the Common Puffball has been translated as 'Wolf's Fart'. In India the warted Mushroom *Bovista pusilla* has been given the name 'Thunder Turd'. An alternative name for the Brown Puffball *Lycoperdon Umbrian* in Mexico is ju'ba'pbich, 'Star Excrement Fungus'. Whether or not these names derive directly from the gastronomic effects of eating Puffballs is usually left unsaid!

FIGURE 32. **Puffball Mushrooms (**Genus *Calvatia*)**.**

Several mushrooms, including Puffballs are known to cause indigestion and flatulence, although several are regarded as edible, assuming they are correctly identified and only ingested in small doses.

Universally, Puffball Mushrooms have a rich history of uses ranging from the practical to the spiritual. For example, indigenous North American peoples such as the Blackfoot tribe endowed smouldering puffball mushrooms *Astraeus hygrometricus* with a celestial significance, as indicated by the Blackfoot name for the mushroom ka-ka-taos, meaning 'fallen stars'. The Blackfoot people believed that puffballs actually represented stars that had fallen to earth as a result of extraterrestrial events that endowed these mushrooms with supernatural properties. In some societies specific reference is made to the planet Jupiter when invoking the extraterrestrial source of puffballs This is a belief that shares some similarities with the mythical history of truffles.

The supernatural properties attributed to puffball mushrooms have included the ability to ward off ghosts and evil spirits, a power that was believed reside in the smoke from smouldering mushrooms once they were set alight (Burk 1983). In addition to mushroom burning as a form of incense, an interesting supplement to this ritualistic activity was the replacement of mushroom spores in the puffball cap with pebbles. This created a spherical rattle that was used by medicine men in ritualistic ceremonies. The reverence for puffballs held by North American tribes is also illustrated by the appearance of painted pictures of these mushrooms around the bases of their teepees. Like several species of Magic Mushrooms, the tendency for some species of Puffball Mushrooms to grow in 'Fairy Circles' has been interpreted as further enhancement of their magical properties by many North American tribes and numerous other cultures that have viewed puffballs as possessing spiritual powers. Further recognition of these magical powers has materialised as lucky charms in the form of necklaces, bandoliers and other ornaments created from dried puffballs. Besides the Blackfoot tribe, other native North American peoples likely to have been familiar with Puffballs and their uses include the Iroquois, the Kiowa, the Menomini (or Menominee), the Ojibwe, the Omaha, the Tewa and the Zuni.

Apart from their spiritual use there is strong evidence that Puffball Mushrooms have practical applications. One example is the use of smoke from burning Puffball Mushrooms to subdue bees to enable honey collection, not unlike modern apiary practices. Smoke from burning mushrooms was also used by North American tribal members to ward off wild animals.

However, probably the most important characteristics associated with puffball mushrooms are their therapeutic qualities. Following archaeological evidence for the Neolithic use of mushrooms for wound dressings and poultices, it is clear that this external use of puffball mushrooms has featured in numerous cultures up to and including the present day.

It is also likely that the ingestion of puffball mushrooms for food and medicine has been part of this history. Particularly over the last few centuries there has been enough evidence to support the edibility of puffball mushrooms. However, their consumption has been confined to the younger, more digestible stage of the mushroom and their restricted intake to safe doses. Subject these conditions immature puffballs may be sliced, diced and added to soups, or dried and crushed before their addition to teas and other drinks. Consumed in this way puffball mushrooms are a source of several bioactive compounds with a broad range of medicinal properties. These include beta-glucans, which are enhancers of the immune system and ergosterol, a steroid alcohol that is important for membrane integrity and has immune-strengthening properties. Puffballs also contain triterpenes, known for anti-inflammatory and antimicrobial properties. Calvacin is a bioactive mucoprotein that was isolated from the giant puffball in the 1960s and investigated as an antitumour agent with a particular focus on lung cancer. (Eroglŭ et al. 2016). Research indicated a delicate balance between remedial and adverse effects, eventually leading to its abandonment as an anticancer agent.

In addition to a mixed bag of remedial properties, puffball mushrooms contain several compounds that are important for healthy growth and general bodily function. These include essential amino acids such as tryptophan, valine, leucine, phenylalanine as well as polyunsaturated fatty acids. It appears, therefore, that these mushrooms may be safely consumed as part of a healthy diet with some remedial benefits, as long as they are consumed in moderation and avoiding toxic lookalikes. Unlike 'magic mushrooms', there is no evidence that puffball mushrooms have been used as inebriants or that they possess any hallucinogenic properties. We can therefore conclude that their historical use was purely dietary or medicinal. The same often cannot be said for other fungi without direct clinical or biochemical evidence.

The Smelliest Fungus.

Another member of the Basidiomycota is the fetid false coral *Thelephora palmata* (FIGURE 33). The species was first described in 1772 by the Italian biologist Giovanni Scopoli, who originally gave it the name *Clavaria palmata.* It was transferred to the genus Thelephora in 1821. It grows in mycorrhizal association with coniferous trees in Europe, North America and Eastern Asia including Fiji and Australia.

It has the dubious reputation as being among the stinkiest of all the fungi, with an odour, some say stench, that has been described as a cross between of rotten cabbage and garlic. The name 'fetid' is clearly well-earned. It has been designated as inedible, at least by humans, although judging from its appalling smell, it's unlikely that many people have tried!

The fungus grows on tree roots and forest floors in small tufts no more than 2 inches high that have been described as coral-like; hence the title 'false coral'.

FIGURE 33. **The Fetid False Coral**. *Thelephora palmata*

The Common Ink Cap.

The common Ink Cap mushroom *Coprinopsis atramentaria* (FIGURE 34.) is widely distributed in temperate climates in Europe and North America where it is commonly found growing on or adjacent to decaying wood.

FIGURE 34. ***Coprinopsis atramentaria*** The Common Ink Cap or Inky Cap mushroom

The fungus was first described by the French biologist Pierre Buillard in 1786 as *Agaricus atramentarius* before being identified as a member of the genus *Coprinus* in 1838 by the Swedish mycologist Elias Magnus Fries. The species name *atramentarius* is derived from the Latin 'atramentum' meaning ink. This refers to the black inky fluid released by the mushroom when it is picked or damaged. The cap is 1-4" in diameter and the stem is 1-4"

Although the Ink Cap has been described as edible in small quantities, larger doses have precipitated several adverse symptoms including nausea, vomiting, palpitations, flushing and a tingling sensation in limbs. These effects are exacerbated if the mushroom is consumed with alcohol, a phenomenon that has earned the mushroom the name 'Tippler's Bane'. In many respects these symptoms mimic the most unpleasant effects of excess alcohol consumption and ensuing hangover and have led to the suggested use of the mushroom as a potential alcohol deterrent. The underlying assumption is that the exaggerated hangover will provide a powerful disincentive for alcohol abuse. Hangover symptoms can occur if even a small amount of alcohol is consumed up to three days after eating the mushrooms, although these effects become milder as more time passes.

The active Ink Cap ingredient responsible for magnifying the adverse consequences of ethyl alcohol (ethanol) consumption is a cyclopropyl glutamine compound called coprine (FIGURE 35.). Its metabolite, 1-aminocyclopropanol, blocks the action of the enzyme, acetaldehyde dehydrogenase, which is responsible for the breakdown of the ethanol intermediate metabolite acetaldehyde in the body. Acetaldehyde is responsible for most symptoms of a hangover which will continue to accumulate if it is not metabolised. Many people might agree that there is a certain skewed logic in creating hypersensitivity to alcohol as a remedy for excess ethanol consumption when the 'cure' by itself (i.e. coprine) may also be responsible for several unpleasant side-effects including potentially serious cardiovascular disorders. The verdict seems to be that the remedial use of the Ink Cap in this way carries too many risks, and that its use should be confined to the role that gave the mushroom its name – ink.

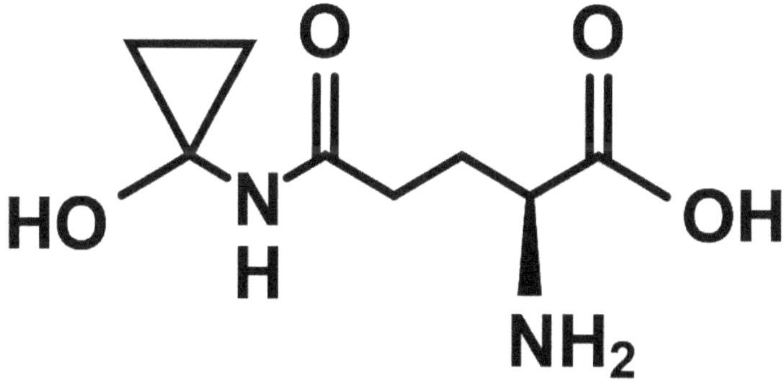

FIGURE 35. **Coprine.** (N^5-(1-hydroxycyclopropyl)-L-glutamine) is the bioactive ingredient of the Common Ink Cap mushroom.

The Role of Mushrooms in History.

Although there is ancient archaeological evidence of an association between mushroom-like fungi and humans based on excavated cookery items and even human remains, the exact nature of this human/fungus relationship remains speculative, even where evidence points to human consumption of mushrooms going back thousands of years. One explanation for the connection between early humans and the fungi could be the possible recreational use of inebriants and other mind-altering substances including psilocybe mushrooms. However, based purely on archaeological evidence it is difficult to differentiate between a recreational 'trip' into an alternative drug-induced world, or participation in a 'spiritually motivated' psychedelic ceremony as practised by several spiritual communities under the influence of magic mushrooms and other hallucinogens. For example, there is ample evidence from Central America going back to at least 10,000 BCE of the involvement of psilocybin mushrooms in ceremonial divination, healing and communing with gods.

The Flesh of the Gods.

In Aztec and Mesoamerican culture mushrooms have long been known as *Teotlnanácatl* "The Flesh of the Gods". Spanish chroniclers of these hallucinogenic practices in Central and South America in the 16th century were both fascinated and horrified by the spiritual role of these Magic Mushrooms in Aztec, Maya, Olmec and Zapotec communities. The use of Psilocybe mushrooms, primarily *Psilocybe Mexicana*, to induce inebriated behaviour was regarded by many Spanish adventurers as blasphemous and an insult to the Catholic church. This resulted in the use of these mushrooms being outlawed by the Spanish conquistadors in 1521 who thereafter imposed Catholicism on the native communities. Consequently, the use of Psilocybe mushrooms went underground and has continued in secret in some communities to this day.

The Aztec emperor Moctesuma II (1466-1520) commonly known as Montezuma was reputed to be a keen consumer of Magic Mushrooms such as Fly Agaric, although he was not a noted drug-taker and was more likely to have used them as an occasional inebriant along with his favourite foods which were chocolate and honey. It is likely that Montezuma also ate Magic Mushrooms as part of the spiritual rituals that alienated the Spanish conquistadors who regarded Aztec ceremonies as decadent.

Some accounts describe Montezuma's attempts at hospitality and reconciliation with his conquerors including sharing his favourite mushrooms. Any discomfort such as diarrhoea or other digestive issues resulting from Spanish overindulgence involving Fly Agaric has since been given the name Montezuma's revenge. This term is still applied today to any desperate unintended visit to the bathroom toilet, whether or not it was caused by Magic Mushrooms. Any attempt at appeasement (or was it revolt?) by the Aztec emperor seemingly ended in failure, as Montezuma is reputed to have met his death at the hands of the conquistadors. However, the identity of the perpetrator(s) of Montezuma's demise remains a matter of debate. Clearly, early exchanges between the imperialists and the native communities were not working, and a different approach was needed. Although, in many respects, large portions of South- and Mesoamerica were was seen as 'The New Spain', a heavy-handed demand for the allegiance of native communities wasn't the answer.

Despite the often-sharp differences between the cultural and religious norms of the Spanish Conquistadors and the native peoples of Meso- and South America, whose territories they occupied, Spanish king Phillip II recognised that aggressive subjugation of the native people wasn't going to work in the long term. The Spanish Conquest of Mexico had taken place between 1519 and 1521, before Phillip's birth in 1527 and his accession to the monarchy in 1556. He was, therefore faced with striking a balance between what, currently, we might refer to as the 'Hawks' and the 'Doves'. Phillip therefore set his viceroys the task of finding out as much as possible about the countries now they were subject to his colonial rule. It was clear that, to establish a harmonious relationship between the Spanish and their colonial subjects, it was important to find out what made these societies work, with particular emphasis on the Aztecs. What were the spiritual and religious influences that were the foundations of this culture? The result of this initiative was a definitive written record of the language and spiritual beliefs of the Aztec people (the Nahua) in the form of a remarkable 12 volume, lavishly illustrated 2,400-page work of art dating back to the late 16th century. It was begun between 1575 and 1579 by the Franciscan friar Bernadino de Sahagún and has since become known as the Florentine Vortex. As the 'Florentine' title signifies, the original work now resides in the form of three books in the Laurentian Library of Florence, Italy, following its subsequent acquisition by the Medici family.

The volume is an examination of the spiritual fabric of the Nahau people and the factors that influenced their religious and spiritual doctrine. For part of the answer, there was no need to look much further than *teonanacatl* the 'flesh of the gods', the hallucinogenic mushroom *Psiolcybin Mexicana*.

Another example of possible tension between the Catholic church and Pagan culture stems from ancient Ireland where an old Gaelic ritual named *Imbas Forosnai* has been handed down largely by word of mouth over several hundred years.

Some insight into the nature of the ritualistic practice can be gleaned from the words Imbas and Forosnai, which can be translated as 'inspiration' and 'illuminated'. The ritual involved the induction of a trance-like state with the aid of psychedelic fungi, which probably included fly agaric (*Amanita muscaria*) as well as psilocybin mushrooms.

In *Sanas Cormaic, Imbas Forosnai* is described as a ritual involving consumption of 'red flesh', which is thought to refer to the red fly agaric mushroom commonly found in Ireland during the autumn months. The *Imbas Forosnai* seems to have been largely maintained as an oral tradition for many hundreds of years in order to preserve its secrecy and help it survive in the face of censorship by the dominant Christian church. However, part of the etymology of the *Imbas Foronai* was committed to the page in the Celtic vernacular dictionary *Sanas Cormaic* in 908 AD during a period of suppression by the church. The provenance of the ritual and its history remains difficult to confirm with certainty, due to the time difference between medieval accounts of the *Imbas Forosnai* and its first appearance. Nevertheless, the antiquity of the *Imbas Forosnai* is confirmed by its mention in perhaps the oldest existing Celtic Irish mythological text, the pre-Christian ancient Irish epic tale *Táin Bó Cúalnge* ("The Cualnge Cattle-Raid"). This suggests that the age of the *Imbas forosnai* can be counted in thousands of years. It is, therefore, reasonable to estimate that the hallucinogenic use of mushrooms in Ireland takes us back at least a few thousand years.

The Fly Agaric Mushroom.

There have been several reports of rock art depicting magic mushrooms from various parts of the world, including claims of Australian cave paintings purporting to show images suggesting that these fungi may have inhabited Australia more than 10,000 years ago. The underlying rationale is that the mushrooms probably migrated from Siberia and eastern Asia to the Australian subcontinent. However, although there is little doubt about the antiquity of Australian cave art and the depiction of an abundance of wildlife in the pictures, the representation of mushroom and other fungi in these paintings and carvings remains highly questionable and the subject of much controversy. Much less contentious is the provenance of rock paintings of hallucinogenic mushrooms in rock paintings from the Algerian Tassili n'Ajjer plateau in the Sahara Desert. Another prehistoric archaeological site featuring 'fungal rock art' is the Selva Pascuala rock shelter, one of more than 20 similar sites celebrating the 6000 year-old ritual use of a fungus that has been identified as psychoactive *Psiolcybe hispanica*.

The fly agaric mushroom (*Amanita muscaria*) (FIGURE 36) is native to many parts of Asia and Europe is thought have migrated from Siberia to Mesoamerica and North America. Fly agaric mushrooms are noted for their hallucinogenic properties and use in shaman rituals. The Hindu god Shiva is frequently portrayed holding a mushroom, signifying the mushroom's spiritual status and we may assume that the fly agaric mushroom has played a part in several ritualistic ceremonies over several millennia. The ancient Hindu text *Rig Veda* which was written by Aryans in the Indus valley between 1500-500 BCE refers to an invigorating extract from a plant called "soma", which is considered to bestow immortality and vitality to consumers. In his book "Soma. Divine Mushroom of Immortality" the ethnomycologist R. Gordon Wasson suggests that the "red plant with no leaves" mentioned in the *Rig Veda* is in fact the fly agaric mushroom.

The name 'fly agaric' stems from its early use as a fly repellent in the days of the botanist Carl Linneus who, in 1753, identified the mushroom as *Agaricus muscarius*, with the species name derived from the Latin for 'fly'. It was later placed in the genus *Amanita*. It is safe to assume that the Fly Agaric mushroom has been prized for its hallucinogenic properties for millennia.

FIGURE 36. **The Fly Agaric mushroom** *Amanita muscaria*

The two natural toxins in the mushroom that are responsible for the mushroom's psychoactivity are muscimol (FIGURE 37) and ibotenic acid (FIGURE 38). Muscimol (FIGURE 37.), also known as pantherine, was first isolated from the Fly Agaric mushroom (*Amanita muscaria*) in 1964. is an alkaloid compound that is chemically similar to the inhibitory neurotransmitter Gamma Amino Butyric Acid (GABA). Muscimol acts as an agonist of the $GABA_A$ receptor thereby stimulating receptor activity. This enhances neuroinhibitory transmission and has a depressive effect on the nervous system resulting in muscle relaxation and a mood-calming effect.

However, the importance of dose is paramount. A history of the use of the mushroom in Siberia indicates that low doses of the mushroom were traditionally used as a stimulant, whereas larger doses were consumed for recreational or spiritual purposes. Reactions to even higher doses fall into the toxic category with symptoms such as nausea, dizziness and nervous twitching. Extreme reactions such as seizures, coma and deaths are rare. All but extremely low doses should be avoided however, subject to regulatory conditions.

The ability of the human body to deal with the more toxic ingredients of the Fly Agaric mushroom is embodied in a procedure that continues to be practiced by Siberian shamans to this day. The process relies on the ability of the liver to filter out some of the more noxious components of the magic mushroom diet, leaving the psychoactive compounds in the urine. The urine is then saved for ritual (re)consumption to prolong or repeat the spiritual experience. In some Arctic Siberian societies, the shamans have opted instead to feed the mushrooms to their reindeer. In such cases the reindeer accomplish the task of concentrating the muscimol in their bladder so that the urine could be drunk for spiritual fulfilment. However, this is not recommended for general use!

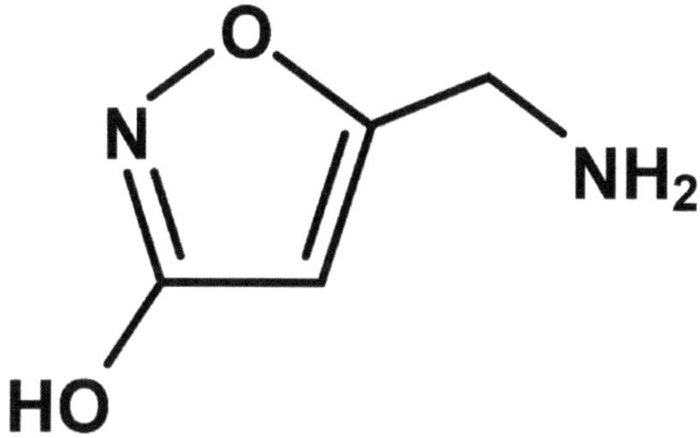

FIGURE 37. **Muscimol** (Pantherine) is a primary psychoactive constituent of the Fly Agaric mushroom (*Amanita muscaria*).

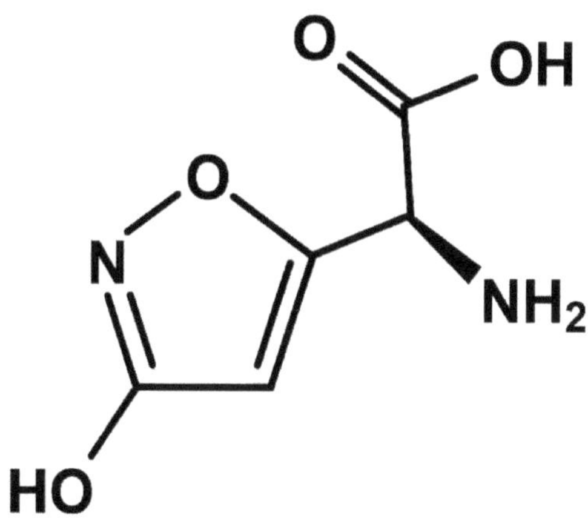

FIGURE 38. **Ibotenic acid**.is an alpha amino acid that is capable of conversion to muscimol.

The Deadly Deathcap.

The most notorious member of the *Amanita* genus is the aptly named Deathcap Mushroom, *Amanita phalloides* (FIGURE 39). This is the most lethal of all the mushrooms belonging to the *Amanita* genus and is suspected of being responsible for 90% of all deaths caused by mushrooms. It is reputed to be the poison of choice used by the Roman empress Agrippina in the murder of her husband, the Roman emperor Claudius. Other notable accidental deaths from Death Cap poisoning include Pope Clement VII in 1534, and the Holy Roman emperor Charles VI of Austria in 1740.

Deadly stories of the Deathcap mushroom continue to this day with examples of both accidental and deliberate poisonings. In 2016 14 serious cases of accidental Deathcap mushroom poisoning were recorded from the San Francisco Bay area, U.S.A. As for deliberate poisonings, in 2023 Australian Erin Patterson was found guilty of the murder of three 'in-laws' by deliberately feeding them Deathcap Mushrooms along with a Beef Wellington dish that she served at a dinner party. She no doubt took advantage of the fact that the toxicity of the deadly ingredient of *Amanita phalloides*, the cyclic peptide α-amantin (FIGURE 40.), is undiminished by cooking.

FIGURE 39. *Amanita phalloides***.** The **Deathcap Mushroom** is the most poisonous of all mushrooms. Its toxin responsible is the cyclic peptide α-amantin (FIGURE 40.).

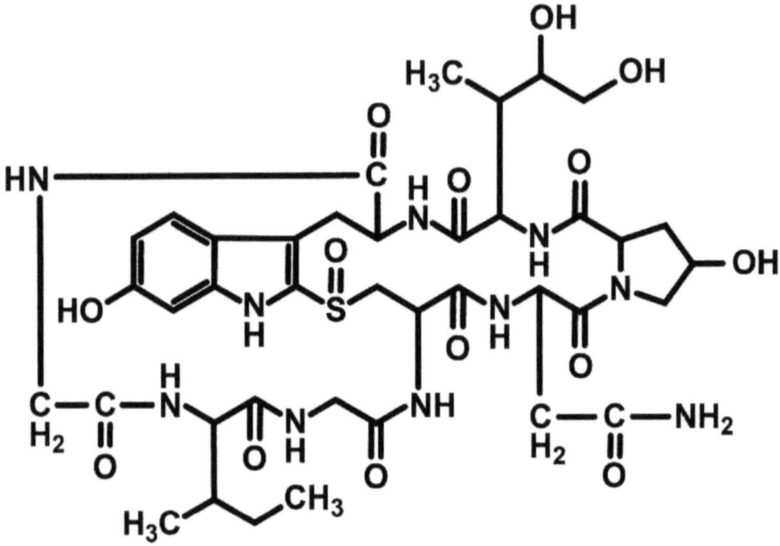

FIGURE 40. **α-Amantin.** The highly toxic ingredient of the deathcap Mushroom *Amanita phalloides*.

There are signs that the landscape regarding magic mushrooms use may be changing. In recent years there has been a re-examination of the potential remedial properties of very low doses of psilocybin and muscalin. This reassessment is based on the relatively recently understood gradation of effects from harmful to beneficial, depending on the amount of mushroom ingested. In other words there is a recognition that, by dismissing the value of these fungi based on the adverse effects of relatively high doses we may have overlooked some of the remedial benefits gained from the ingestion of these mushrooms and their products at much lower doses.

To some degree we can trace this reassessment of magic mushrooms to research performed by the Swiss pharmaceutical chemist Albert Hofmann, who first isolated psilocybin from the psychedelic mushroom *Psylocybe Mexicana* (FIGURE 41.) in the late 1950s. As a result of Hofmann's work, psilocybin was first synthesised in pure form in 1958 and different structural analogues of the compound were subjected to tests of their relative psychoactive properties. In other words, these active ingredients of magic mushrooms were subjected to an intensive study of Structure Activity Relationships (SAR). As the name implies SAR represent a branch of toxicology dedicated to understanding the relationship between chemical structures and their physiological/biochemical function (Wright 2025). In what might be described as a 'new twist' to this approach, Hofmann was willing to ingest these psychedelic compounds himself in order to assess their hallucinogenic properties. As the discoverer of the psychedelic effects of the hallucinogen LSD, it would have been difficult to find a better qualified researcher than Hofmann for this topic!

FIGURE 41. ***Psilocybe mexicana.*** This mushroom provided the original source of psychoactive compounds Psilocybin and Psilocin isolated by Albert Hofmann.

Research projects such as these proved pivotal in our understanding of both chemical structure and dose in determining optimal bioactivity. We are reminded of the teachings the Swiss-German alchemist and philosopher Paracelsus (1493-1541) who was primarily responsible for introducing the concept of dose into the study of natural toxins. Paracelsus is famous for the quote:

"All things are poisons, for there is nothing without poisonous qualities. It is only the dose which makes a thing poisonous".

Since the time of Paracelsus nearly 500 years ago this philosophy has been applied to not only natural toxins but also prepared chemical extracts and newly manufactured chemical compounds. In recent years the concept of dose has guided our thinking and has led to a reassessment of some bioactive substances whose effects range from toxic to beneficial depending on the amount ingested. This has resulted in a more nuanced attitude to several natural toxins, including some that are used as recreational drugs. For example muscimol and *Amanita muscaria* mushrooms have rarely been used recreationally. However, by the mid-2020s, there have been signs of a marked increase in the ingestion of these substances for both recreational use and medicinal benefit. Recent years have seen an increase in the term 'microdosing' when referring to portions of magic mushroom constituents that may be safely prescribed for therapeutic benefits. Also there has been an increasing involvement of scientific research in defining safe doses. Commonly cited examples include the therapeutic use of Fly Agaric and similar bioactive mushrooms is as sedatives, relaxants and as agents to improve sleep. However, currently recognised therapeutic benefits associated with mushrooms also include relief from depression, anxiety and treatment of PTSD[20] and addictive behaviour.

This shift in attitude to Fly Agaric and other bioactive mushrooms is reflected by changes in regional and national legal legislation governing the use of Magic Mushrooms and their constituents. For example we now see a trend towards the decriminalisation of psilocybin and similar psychedelic compounds in numerous countries. As of 2024, some psychedelics have been legalised or decriminalised to varying degrees in South Africa, Australia, the Bahamas, Brazil, the British Virgin Islands, Canada, parts of Central Africa, the Czech Republic, Costa Rica, Israel, Jamaica, Mexico, Nepal, the Netherlands, Panama, Peru, Portugal, Samoa, South Africa Spain, Switzerland and the United States.

However an examination of the governing jurisdiction in these countries reveals many ambiguities showing that the legality of these compounds is still far from settled. The confusing state of international legislation is illustrated by an examination of the positions of various countries on this issue.

In the U.K. psilocybin mushrooms were made illegal under section 21 of the (2005) U.K. Drugs Act. This act declared raw or prepared psilocybin mushrooms as a class A drug alongside cocaine, heroin, morphine, oxycodone, MDMA[21] ('Ecstasy'), methamphetamine, opium, LSD, hydrocodone, DMT and mescaline[22].

---

[20] PTSD is post-traumatic stress disorder.

[21] MDMA is 3,4-Methylenedioxymethamphetamine (known as 'Ecstasy') was introduced as a psychotherapy agent in the the 1970s, but became popular as a recreational drug in the 1980s.

[22] Class A drugs are defined as having no recognised medicinal benefits. However, in view of updated information on the potential therapeutic qualities of psilocybin, this is subject to potential change and re-classification.

Possession and use of psilocybin and psilocin has been prohibited since the 2005 U.K. Drugs Act. However, mushroom spores, which do not contain psilocybin, are not regulated. 'Grow kits' can be bought legally through selected sources because they do not contain psilocybin and psilocin. The legal justification of these 'kits' containing spores is often expressed in terms of their 'microscopic interest', seemingly circumventing a potential ban on the cultivation of the psilocybin mushrooms themselves in favour of carrying out further research that might have a bearing on the future therapeutic benefit of the mushrooms and their bioactive constituents. Other countries such as Japan, Poland, Russia, Sweden, Sri Lanka, Turkey and Ukraine apparently have adopted a similar, if somewhat equivocal attitude to the mushroom spores, although it is not easy to find two countries with identical legislation concerning magic mushrooms and their relatives. In fact, in many cases regulations from different countries may seem to be contradictory.

At one end of the scale, we find Jamaica where psilocybin mushroom have never been illegal, and where there is an open market for mushrooms that are freely cultivated and sold. At the other end of the scale Psilocybin mushrooms and their products are completely illegal in Hong Kong, Hungary and Iceland. Although the possession, marketing, transport and cultivation of Psilocybin mushrooms are illegal in India, prohibition is rarely enforced. In many cases the penalisation of magic mushroom use is regarded with indifference. For example, in a 2025 high court case in the Indian coastal state of Kerala it was declared that magic mushrooms were neither narcotics nor psychotropic substances, and therefore not subject to legislation. Many areas of India include psilocybin and psilocin among psychotropic drugs forming the basis for a lucrative underground psychedelic tourist trade.

In the U.S. psilocybin was initially classed as schedule 1 substance under controlled substances act of 1970 and most U.S. states subsequently regarded magic mushrooms simply as a source of illicit drugs and declared them illegal. This blanket categorisation has since been subject to what might be called 'selective enforcement' among U.S. states and U.S. cities, and it is not unusual to find the words "illegal, but unenforced" attached to the laws and regulations involved.

The current picture regarding international legislation is largely divided into four ruling categories, namely 'Possession', 'Sale', 'Transport' and 'Cultivation', and the regulations governing psilocybin and similar psychedelics are generally organised around these activities. Legislation and regulations remain equivocal and may differ from state to state and even from city to city. For example, in 2021 the state of Oregon legalised both psilocybin and psilocin for mental health treatment. However, selling psilocybin without being licensed may still attract fines or imprisonment. In 2022 Colorado decriminalized psilocybin including psilocybin mushrooms, and psilocin, together with DMT (dimethyltryptamine) and other mind-altering drugs for those aged 21 and over.

This legislation also legalized licensed "Healing Centers" where patients can receive psilocybin under supervision (which may also be expanded to include DMT, ibogaine, and mescaline). On reaching the age of 21 people are allowed to cultivate, transport and possess these substances but not sell them for personal use. Other jurisdictions in the United States where psilocybin mushrooms are decriminalized include Ann Arbor and Detroit, Michigan; Oakland and Santa Cruz, California; Easthampton, Somerville, Northampton, and Cambridge, Massachusetts; Seattle, Washington; and Washington, DC.

A review of legislation covering the use, cultivation and distribution psilocybin mushrooms often makes for confusing reading and remains a 'moving target' in many respects. In future, much will depend on striking the appropriate balance between the potentially harmful and remedial properties of these fungi and their relatives as more is learned about their pharmacological and psychoactive properties.

The Impudent Phallus.

Phallus impudicus (FIGURE 42.) is an edible mushroom of the Phallaceae family commonly found in Europe and North America where it typically inhabits woodland floors. It also has a mycorrhizal association with the root systems of some trees. Its common name, stinkhorn, reflects its two most characteristic features, its shape and pungent smell. The sickly odour given off by the mushroom is similar to rotting meat and serves to attract flies and other insects to the sticky mass of spores on the mushroom's cap. Unlike many other mushroom species, stinkhorns rely on insects rather than wind to disperse their spores. The unpleasant smell of the fruiting body can be removed by cooking, producing a mushroom that is commonly and safely eaten.

It is difficult to ignore a mushroom whose Latin name is translated as shameless penis and much has been made of its phallic shape in folklore. Probably because of its phallic shape, the stinkhorn has gained a long and colourful reputation over several hundred years In his *General Historie of Plants* published in 1597 the botanist John Gerard referred to the stinkhorn called it the "pricke mushroom" and there have been several other similar references. However, the English herbalist John Parkinson was somewhat more refined in his nomenclature, referred to it as "Hollanders workingtoole" or "*phallus hollandicus*" in his encyclopaedic *Theatrum botanicum* of 1640. Although *Phallus impudicus* is probably the best known and most notorious of Phallaceae family, more than 20 varieties of this mushroom have been given the name stinkhorn.

Phallus impudicus (FIGURE 42.) sprouts from an egg-shaped base. It consists of a 'rubbery' stalk and a sticky cap, known as the 'gleba'. The gleba has a pungent odour that attracts flies and other insects that are responsible for dispersal of the spores for reproduction.

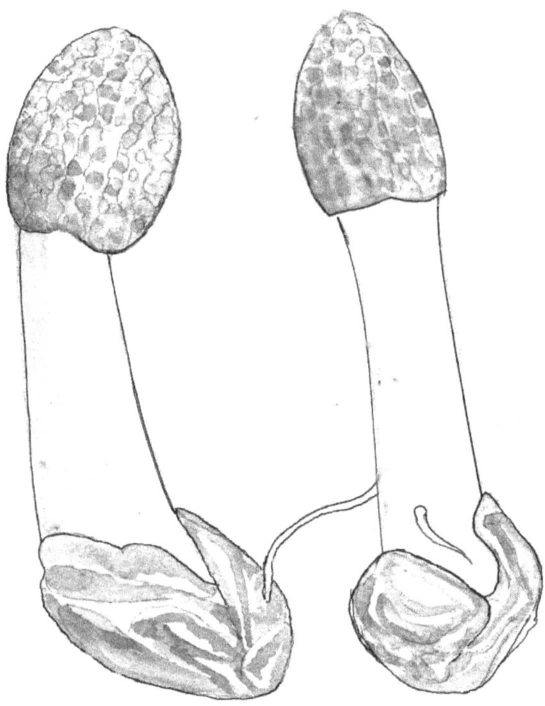

FIGURE 42. **The Impudent Phallus Mushroom**. *Phallus impudicus*

As a result of their phallic shape it is, perhaps, no surprise that the stinkhorn mushrooms are part of a folklore narrative that often features sex. This includes a rich history of the mushroom as a reputed aphrodisiac. Stories of sexual desire and their fulfilment through ingestion of stinkhorn mushrooms abound in many different cultures where stinkhorn mushrooms are common. Early Hawaiian narratives on the sexual properties of the mushroom contend that simply physical contact with the stinkhorn may be sufficient to achieve sexual satisfaction. However, these accounts state that it is only women who are able to achieve sexual fulfilment in this way. Apparently, no such luck for men!

The aphrodisiac properties of stinkhorn mushrooms were also a part of ancient Roman folklore, which featured the incorporation of magic mushrooms into an elixir that was reputed to boost the sex-life of both men and women. We should be reminded, however, that there is no solid experimental or pharmacological evidence for any aphrodisiac properties attached to stinkhorn mushrooms: nothing beyond the power of suggestion perhaps?

However, the power of suggestion is not always well received! Several societies have regarded the stinkhorn mushroom as obscene because of its shape and in some religious circles the names "Devil's Horn", "Devil's Stinkpot" and "Satan's Member" speak for themselves.

Despite the enthusiasm for stinkhorn family's supposed aphrodisiac qualities in many parts of the globe, the idea of a mushroom resembling an erect penis as a cultural icon does not go down well in some circles. Clearly Charles Darwin's daughter, Etty, could be counted among this disapproving group. Writing about life in Victorian Cambridge, Charles Darwin's granddaughter Gwen Raverat describes a 'sport' organised by her mother Etty in which woodland stinkhorn mushrooms were hunted down by their rancid smell, speared with a pointed stick and gathered by the basketful for surreptitious burning on the drawing room fire. The burning was conducted in strict secrecy in order avoid the corruption of household staff who might otherwise be exposed to these obscene mushrooms. However, not all members of the Phallaceae family have been met with such disgust.

The Veiled Lady.

A particularly attractive member of the stinkhorn family of mushrooms, *Phallus luteus* (FIGURE 43), has a much more 'delicate' cultural history. The mushroom is widely distributed in Eastern Asia and in several tropical regions including Mexico and has received numerous accolades as an object of beauty, with many references to the yellowish lacy 'veil' that is responsible for the Latin adjective *indūsĭātus*, (translated as 'wearing an undergarment') that provided the species' original name *Phallus indūsĭātus*,. Among the common names for this stinkhorn variety are long net stinkhorn, crinoline stinkhorn, basket stinkhorn, bridal veil fungus and veiled lady. Some people might find it amusing that, simply by adding a veil that the mushroom suddenly becomes a lady even though the generic name remains *Phallus*! During most of the 19th century this member of the stinkhorn mushroom family underwent a series of taxonomic name changes. For example, it was initially described in 1936 and was named *Dictyophora lutea*. However, it was returned to the genus *Phallus* in 2008 and given the species name *Phallus luteus*. Other species of *Phallus* also appear in the literature, although their provenance remains a topic of discussion. In his 1798 dissertation on the *Phallus* genus the French botanist Étienne Ventenat makes reference to the species *Phallus hadriani*, a mushroom similar to *Phallus impudicus*, but with a more thimble-shaped cap. However, some have since speculated that Ventenat's discovery was simply of an immature *P. impudicus*. This will probably remain a mystery. The species name *hadriani* is named after the Dutch botanist Hadrianus Junius (1512–1575).

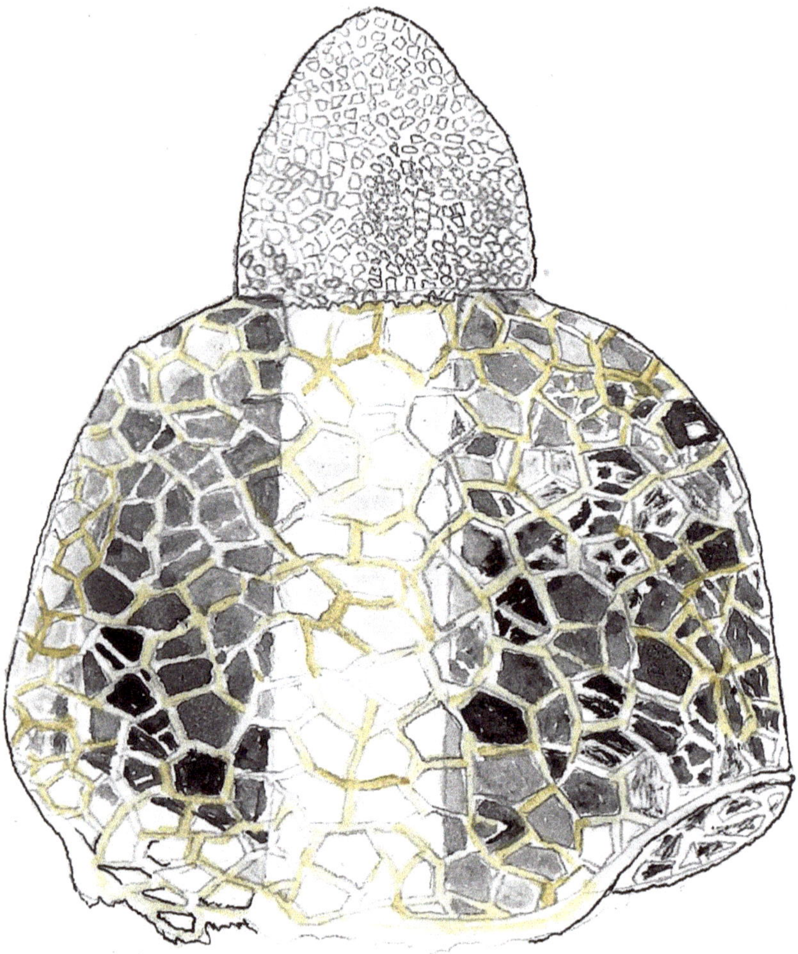

FIGURE 43. *Phallus luteus.* A stinkhorn mushroom known as '**The Veiled Lady**'.

The Octopus Stinkhorn.

The Octopus Stinkhorn (*Clathrus archeri*), otherwise known as the Octopus Mushroom or the Devil's Fingers is another member of the stinkhorn family that is widespread globally, although it was first described from Tansmania in 1980 by British mycologist Donald Dring.

This stinkhorn variety is characterised by the putrid smell that is common to the stinkhorn mushroom group. However, unlike several foregoing examples, any resemblance to a phallus is missing from this species. Instead, as we follow the maturation of this mushroom (FIGURE 44.) we see a somewhat bizarre sequence of images as the fungus hatches from its egg-like sac in a form that resembles a red octopus. The more pedantic among us will note, however, that the fungus has between 4 and 7 arms rather than the eight possessed by an octopus. Nevertheless the 'suckers' on the arms of the fungus have a certain authenticity about them. The individual arms of the Octopus Fungus can each grow up to about 4.5" in length, which means that a mature specimen is about the size of an adult human hand. Some observers have pointed out that the appearance of a fully grown specimen is closer to a starfish than an octopus. The "egg" of the Octopus Fungus, has been described as non-toxic and edible. However, it is rarely consumed because of its unpleasant taste.

FIGURE 44. **The Octopus Stinkhorn Mushroom.**
Pictures by Anna Counts.

Stinky Starfish.

Another member of the stinkhorn group bearing a strong resemblance to the octopus stinkhorn is the anemone stinkhorn *Aseroë rubra* (FIGURE 45.). To further confuse matters, both of these stinkhorn varieties sometimes share the alternative common name starfish fungus.

Despite having common names belonging to marine organisms such as anemones, octopuses and starfish, these stinkhorn variants are strictly woodland and grassland inhabitants. In addition to similarities in their appearance, they are also candidates for the title of worst smelling fungus. Their pungent aroma serves to attract flies and other insects as dispersal agents for reproductive spores. The faecal/carrion smell of *A. rubra*.is reflected in its generic name *Aseroë* which is an amalgum of ancient Greek words meaning 'disgust' and 'slime'. The species name 'rubra' refers to the bright red colour of the fungus.

Aseroë rubra was initially discovered in Tasmania in 1792 by French botanist Jacques Labillardière and became the first indigenous Australian fungus to be formally identified. It remains as a native of Australia, New Zealand, but has since been exported to several other parts of the world including Asia and North America.

FIGURE 45. *Aseroë rubra*. The **Starfish Fungus** or Anemone Stinkhorn (courtesy Casliber). Its pungent smell is a powerful attractant for insects that assist with spore dispersal.

The Lion's Mane.

The Lion's Mane fungus *Hericium erinacius* (FIGURE 46) is native to Europe, Asia and North America where it is often found growing in clumps on (usually dead or decaying) hardwood trees such as beech and the North American maple. Throughout its extensive range it has several names, several forms and several uses. Although it is often referred to as an edible mushroom, its appearance might not fit most people's idea of a mushroom. There are, in fact, several species of the *Hericium* genus that contribute to the varied appearance of the fungus, and common names for this group also reflect its heteromorphic nature. In addition to Lion's Mane, we also find names such as the Bearded hedgehog, and Bearded tooth fungus. References to a beard or mane reflect the 'hairy' fibres that dangle from the fruiting body. However, different specimens of the Lion's Mane fungus can diverge markedly in appearance. As shown in FIGURE 45, the white-bearded characteristic of some seems to be replaced by a cauliflower-like appearance in others. Quite a comedown for the King of the Jungle!

FIGURE 46. **Lion's Mane.** 'King of the Jungle? Or just a 'Common All Garden' vegetable like a cauliflower?

Throughout its extensive range it has several names, several forms and several uses. Although it is often referred to as an edible mushroom, its appearance might not fit most people's idea of a mushroom. There are, in fact, several species of the *Hericium* genus that contribute to the varied appearance of the fungus, and common names for this group also reflect its heteromorphic nature. In addition to Lion's Mane, we also find names such as the Bearded hedgehog, and Bearded tooth fungus. However, different specimens of the Lion's Mane fungus can diverge markedly in appearance.

In addition to the numerous names for the Lion's Mane Fungus that are based on its physical appearance, other common names for the fungus simply reflect their geographical distribution or origin. For example, the Japanese name for the fungus, Yamabushitake, is translated as 'Mushroom of the Mountain People', which refers to a religious sect noted for their spiritual practices. The Lion's Mane fungus itself is not noted for its spiritual powers, and its early Chinese origin as a medicine was reputedly based on its ability to improve mental sharpness and clear thinking rather than the promotion of psychedelic activity. The potential for Lion's Mane fungus to enhance clarity of thought remains a topic of interest many centuries later, although scientific information remains sparse. Published data largely result from *in vitro* experiments rather than controlled clinical investigations. Japanese studies by Shimbo *et al.* (2005) and Mori *et al.* (2008) on cultured neural cells from rats and humans respectively both report the stimulatory effect of *Hericium erinaceous* on what are described as 'nerve growth factor' content. Shimbo *et al.* (2005) also document an increase in the neurotransmitter catecholamine following the exposure of rat brain cells to the fungus.

As a footnote, it is illegal to harvest Lion's Mane fungi 'in the wild' under the 1981 Wildlife and Countryside Act. This has nothing to do with its medicinal or pharmacological properties, as is simply because of its classification as an endangered species.

Shitake Mushroom.

Another popular edible mushroom of eastern Asian origin is the Shitake mushroom (FIGURE 47), *Lentinula elodes*. The Japanese name for the mushroom, 'shitake', is derived from its most favoured host, a beech tree known in Japan as 'Shi', in combination with the Japanese term for mushroom, 'take'. Other hardwood hosts include oak, maple and chestnut. The mushroom has been part of Chinese and Japanese folklore for many hundreds of years during which time it has gained a reputation as a culinary staple with some medicinal benefits. Its reputation in China and Japan has extended to treatment of high blood pressure, high cholesterol and augmentation of the immune system. Its stimulatory effect on the immune system has resulted in its supplementary role in the treatment of some forms of cancer, including stomach cancer. The success of Shitake mushrooms in the treatment of hypertension and the control of blood cholesterol levels seems to be related to the presence of β-glucans, which provide a metabolic substrate for microbiota. Byproducts from this metabolic activity include short-chain fatty acids which have a number of positive effects on several physiological activities including cardiovascular activity, oxidative stress and immune function.

Despite the medicinal reputation of the Shitake mushroom its primary use worldwide is as a delicacy. This has led to its global cultivation being perhaps second only to the common button mushroom *Agaricus bisporus*. Writings from Longquan County, China in 1209 record what appear to be the first attempts to culture Shitake Mushrooms by placing logs and other wood debris in the vicinity of existing mushroom colonies. Infestation can be increased by drilling holes in existing bark and logs. More than twenty 'crops' can be cultivated in this way over a period of at least five years.

FIGURE 47. **Shitake Mushroom.** *Lentinula elodes*, also known as the Black Forest Mushroom, Golden Oak Mushroom and Sawtooth.

The Bleeding Bonnet.

The Burgundydrop Bonnet mushroom was initially identified as *Agaricus Haematopus* by South African mycologist Christiaan Persoon in 1799 but was moved to the genus Mycena in 1871. *Mycena haematopus* (FIGURE 48) is also known as 'Bleeding Mycena', 'Bleeding Bellcap or the 'Bleeding Fairy Helmet' is one of many species of fungus belonging to the genus *Mycena* in the family Mycenaceae. Members of this genus are to be found throughout the globe and are common in Europe and North America, Venezuela & Japan where they grow in clumps on the rotting bark and fallen branches of hardwood trees such as oak and beech. Individual mushrooms consist of pinkish brown caps approximately 1" in diameter with thin stems as long as 4".

The name 'bleeding' that often appears in the common name for these mushrooms is derived from their exudation of a blood red latex from parts of the stem and/or cap that sustain breakage or damage (FIGURE 49). The term bleeding has been incorporated into the species name *haematopus,* which includes the word 'Haem', which is Latin for blood. In the closely related species *M. Lacryma* another Latin derivative 'La cry ma' *vis.* 'crying' is employed to describe this phenomenon.

FIGURE 48. **The Burgundydrop Bonnet**, *Mycena haematopus)* also known as 'bleeding Mycena'

FIGURE 49. **Bleeding Mycena.**

Mycena fungi are usually found growing in clusters on the rotting trunks and branches on dead or dying deciduous trees. However some *Mycena* species may be discovered in more coniferous woodland growing on bark and pine needle beds. Like most fungi belonging to this genus, the Burgundydrop Bonnet mushroom is considered to be saprotrophic, i.e. it uses a mixture of chemical compounds, primarily alkaloids to extract nutrients from decaying wood. Whether or not the fungi actually contribute to the death of the host trees is still a matter of conjecture.

The fungus was initially named *Agaricus haematopus* by the South African mycologist Christiaan Persoon in 1799 but was incorporated into the genus *Mycena* in 1871. This genus currently comprises about 500 species of fungi of which *Mycena haematopus* is probably the best known. Most individual fruiting bodies of the *Mycena* genus are quite small and fragile. *M. haematopus,* one of the largest of the genus, is characterised by a bell-shaped cap with corrugated edges, usually no more than 4 cm in diameter and a thin fragile stem that is usually between 4-9 cm. in length. The caps are usually pale brown in colour but often have a red/pink tinge.

Mycena species, but not all, have been described as 'edible' although most are regarded as too small to be considered as food. Although *M. Haematopus* is not regarded as a significant source of psychoactive compounds, some species of *Mycena,* such as *M. pura* contain small amounts of the hallucinogenic alkaloid muscarine (also found in *Amanita*), which is capable of inducing convulsions and even death at high concentrations. The general consensus is that the Mycena species are generally too insubstantial to be regarded as edible, but risky in higher quantities. This somewhat contradictory stance suggests that the safer specimens of these mushrooms are too easily confused with toxic species. The conclusion is that they are probably best left alone!

Some *Mycena* species including *Mycena haematopus* have been described as weakly bioluminescent, although investigators have often needed to manipulate the wavelength of the light source in order to detect this phenomenon. Because of the difficulty in observing the mushroom's bioluminescence, some have cast doubt on its value to the mushroom. However, others have argued that bioluminescence of the mushroom may be a means of improving its visibility under low light conditions to attract animals such as insects and other organisms that might aid in spore dispersal and reproduction. After all, perhaps some of these airborne dispersal agents may be better able to detect bioluminescence than any human observer!

The truth seems to be that bioluminescence does exist in some species of *Mycena* but is subject to considerable interspecies variability. For example, *Mycena lux-coeli*, the 'heavenly light mushroom' native to forested mountainous regions of Japan, takes its name from its ability to glow in the dark.

However, bioluminescence is not unique to the *Mycena* genus and may, indeed, be a phenomenon that can be traced back to ancient history. Over 2000 years ago Aristotle spoke of the "cold fire" of fungi that lit up the forest at night. Currently over 70 species of fungus are known to be capable of luminescence including several members of the *Mycena, Panellus* and *Armillaria* genera. The name 'Foxfire' is often given to fungi displaying luminescent characteristic

A noted example is the so-called 'Bitter Oyster fungus'. This species was originally identified as *Agaricus stipticus* in 1783 by the French mycologist Jean Buillard but was given the generic name *Panellus* in 1879. Interestingly, the luminescent form of this fungus is only found in eastern North America, although its global distribution covers Eurasia, Australia as well as Canada and the U.S.A. Bioluminescence is embodied in the name Jack O' Lantern mushroom *Omphalatus olearius* that was originally discovered growing on timbers that were being used to support mining activity in North America.

Ear to the Ground.

The Jelly Ear fungus, also known as the Wood Ear fungus, *Auricularia auricula-judae* (FIGURE 50) is a brown saprotrophic wood-rotting fungus native to Eastern Asia although the fungus can be found throughout the year in Europe, where it normally grows on wood of bushes and broadleaf trees with a preference for the alder tree. The Jelly Ear fungus has a long medical and culinary history stemming from Chinese folklore and has been used in folk medicine in Europe as recently as the 19th century for the treatment of a several medical conditions including jaundice, sore throat and sore eyes. It has also been used as an astringent in several applications.

The fungus also possesses a complex, somewhat controversial narrative surrounding its (several) vernacular names. Also known as *Auricularia auricula*, it has also been described from European sources. However, the Chinese and European versions of the fungus have been shown to have genetic differences, and as many as seven closely related species my be responsible for a certain amount of taxonomic confusion.

The fungus was initially described as *Tremella auricula* by Carl Linnaeus in 1753 but was renamed *Tremella auricula* in 1789 by French biologist Pierre Buillard, with the generic name *Tremella* reflecting the jelly-like, trembling nature of the mushroom cap. The species was given the name *Auricularia auricula-judae* in 1888 by German mycologist and military doctor Joseph Schröter. Although the generic name

Auricularia and the species stem *auricula* clearly refer to the ear-shaped fungal cap the species name *auricula-judae* has proved to be more contentious. The name *auricula-judae* was criticised by American mycologist Curtis Lloyd as an antisemitic epithet *vis.* Jew Ear. However, it has been postulated that the '*judae*' portion of the species name has a much more specific, biblical origin. In this version of the species name *judae* is considered to refer to Judas Iscariot, who is reputed to have betrayed Jesus Christ for the sum of thirty pieces of silver. He is then reputed to have committed suicide by hanging himself from an alder tree, which happens to be the favourite host of the jelly-ear fungus. Whichever side of the narrative you prefer, the somewhat cumbersome species name *Auricularia auricula-judae* has been mostly retained, although the name *Auricularia auricula* still appears in some publications.

FIGURE 50. *Auricularia auricula-judae.* **The Jelly Ear Fungus** also known as the Wood Ear Fungus is parasite and saprophyte that grows on (mainly alder) trees in Asia and Europe.

Green Elf Cap.

Chlorociboria aeruginascens (FIGURE 51) and the very similar *C. aeruginosa* are members of the Chlorociboria genus in the Phylum Ascomycota. (Class: Leoteomycetes, Order Helotiales) that are found in the British Isles. However, more than 20 species of this genus have been identified in temperate regions of Europe, North America and Asia in the Northern Hemisphere. The fungi are saprobic and found growing in clusters on the dead wood of deciduous trees such as birch, hazel and especially oak. Their caps can grow to about 2" in diameter but are often much smaller. The stem of the fungus is usually no more than 1.5" tall. Despite their small size, they are easily recognised by the bright turquoise colour of their fruiting bodies. Their distinctive shape and colour have resulted in the common names Green Elf Cap and Emerald Elf Cap.

The dye from these fungi is responsible for staining their wood substrate a turquoise green colour (FIGURE 51.); a phenomenon that has resulted in the name 'Green Oak' for wood that has been stained by the fungus. Although this blue-green staining may detract from the value of the wood as a furniture constituent, it has opened up a new, and highly lucrative niche for the blue-green wood.

FIGURE 51. *Chlorociboria aeruginascens.* Known as **Emerald Elf Fungus** or Green Elf Fungus.

Despite not being used for the manufacture of major household furniture the turquoise wood is greatly prized as a decorative item that can be incorporated into mosaic images and patterns embellishing small boxes, chests and wooden panels. Tunbridge Ware jewellery boxes emanating from the Kentish town of Tunbridge Wells are among several items where the blue-green wooden shapes may become part of the inlaid designs that adorn the wooden veneer surface. Such designs often command high prices.

FIGURE 52. **The Emerald- (or Green-) Elf Fungus** stains its wood substrate with a blue-green dye. Photo by Anna Counts.

The pigment responsible for the blue-green colour of the Emerald- or Green Elf Cap mushroom is the dimeric[23] quinone xylindein (FIGURE 53).

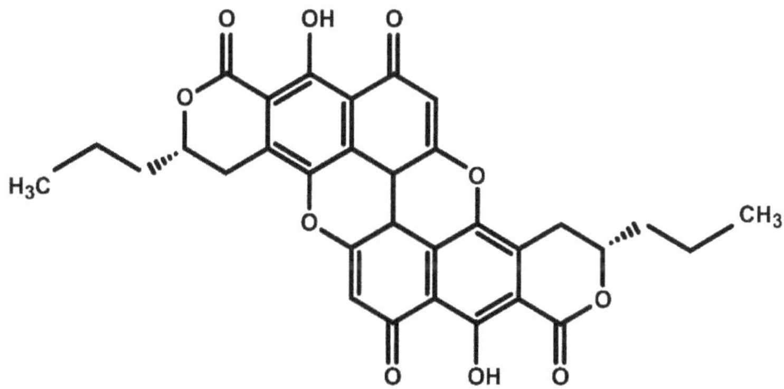

FIGURE 53. **Xylindein** A quinone responsible for the blue- green colour of the Green- or Emerald Elf fungus. 8,16-dihydroxy-3,11-dipropyl-3,4,11,12-tetrahydro-pyrano [4,3-h] pyrano [4',3';5,6] xantheno [2,1,9,8-klmna] xanthene-1,7,9,15-tetraone.

Pochonia. Worm Killer.

Another member of the fungal division Ascomycota occupying a particularly intriguing position as a source of bioactive chemical components is the saprophytic fungus *Pochonia chlamydosporia* (FIGURE 54.). This fungus was first described by the Mycologist H. N. Goddard in 1913 as a

[23] A dimeric compound has a 'twin' structure composed if two similar, joined units.

member of the genus *Verticilium* in 1913, but has since been ascribed to the genus *Pochonia*. The generic name *Pochonia* is a reference to the French microbiologigist Jacques Pochon, an expert on soil microbiology.

P. chlamydosporia is known as a symbiont of the root system of vegetable and crop plants such as lettuce, tomato and wheat where some of its bioactive chemical components enhance the growth of host plants. However its chief claim to fame is its universal role as a nematocide. Although most nematode worms are quite harmless they also include species that are highly damaging parasites of plants throughout the world. Therefore, apart from any growth enhancing properties of the fungus itself, its ability to eliminate harmful nematode parasites will also improve crop yield. Major agricultural pests include the nematode genera *Heterodera* and *Globodera*.

Some nematodes such as *Ascaris* also parasitise animals including humans where they cause gastrointestinal problems. Neurological diseases and blindness in humans may also result from the migration of some parasitic nematodes from the digestive system to other body tissues. Human infection by nematodes may be caused by the ingestion of contaminated food and other matter resulting from poor hygiene. However some nematodes such as hookworms and threadworms are capable of parasitising animal via skin penetration. Diseases caused by these nematodes affect hundreds of millions of humans annually. It is, therefore, obvious that an effective nematocide is a major asset to the international agricultural industry as well as reducing some risks of human infection, particularly in tropical communities where nematode infestation thrives.

Efforts to deploy *Pochonia chlamydosporia* as a nematocidal agent have resulted in the development of pochonia-based pesticides designed to reduce soil nematode numbers. The mode of action of the fungus was originally described by Wilcox and Tribe (1974) as a lethal parasitism of nematode eggs following the breakdown and digestion of the eggshell by fungal extracellular enzymes. Different strains of *P. chlamydosporia* are cultured for this purpose on a commercial scale and may take the form of a mycelial suspension applied to nematode-infected plants or the release of reproductive 'chlamydiospores' in habitats known to be infected with nematodes. Commercial production of *P. chlamydosporia* for this purpose now takes place in North and South America, Europe and China.

Various bioactive products of *P. chlamydosporia* have now been identified, including secondary metabolites such as aurovertins. Aurovertins, which were originally isolated from the penicillium fungus, are a group of over 40 compounds largely consisting of chains of 2-C acetate units. These heteropolycyclic molecules, classified as polyketides, have strong antitumour properties resulting from their role as inhibitors of ATP synthesis. Inhibition of ATP synthesis and hydrolysis also contributes to their toxicity to a broad range of animals and bacteria. These properties underpin the widespread culture and use of *P. chlamydosporia* as a biocontrol agent. Although its primary use is for nematode control, it has also been employed as a more broad spectrum vermicide, insecticide and molluscicide.

FIGURE 54. ***Pochonia chlamydosporia.*** A worm-killing fungus that lives on the root systems of vegetables and crop plants.

Ketamine. Old Drug or New Pesticide?

An intriguing member of the chemical armory of *P. chlamydosporia* is a bioactive component that is thought to act as a defence against consumption of the fungus by insects and roundworms, as well as possessing nematocidal properties. This bicyclic compound, Ketamine, has a rich human history that is far removed from its roles as a fungal nematocide or more general pesticde. The name ketamine is derived from the compound's ketone structure characterised by the presence of a carbonyl (C=O) group. Ketamine belongs to a class of compounds known as cyclohexanones whose basic structure consists of a hexagon ring containing a double-bonded oxygen atom (FIGURE 55).

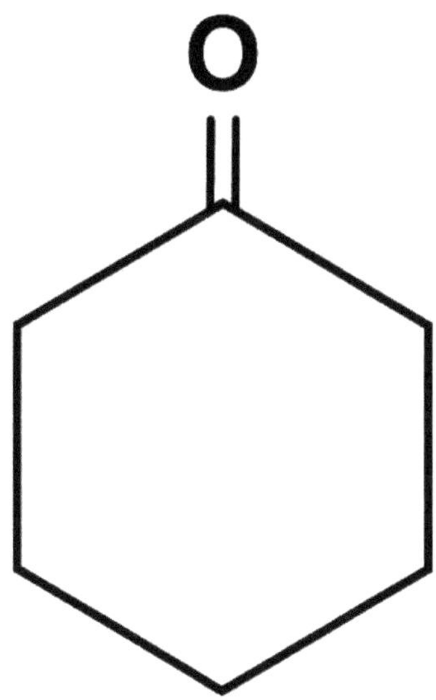

FIGURE 55. The **cyclohexanone** group forms the basis for the drug Ketamine.

In the ketamine cyclohexanone derivative one of the hydrogens at position 2 is substituted by a 2-chlorophenyl group, while the other is substituted by a methylamino group (FIGURE 56). The result is an extraordinarily versatile compound capable of interacting with a variety of physiologically active sites including those of the Central Nervous System.

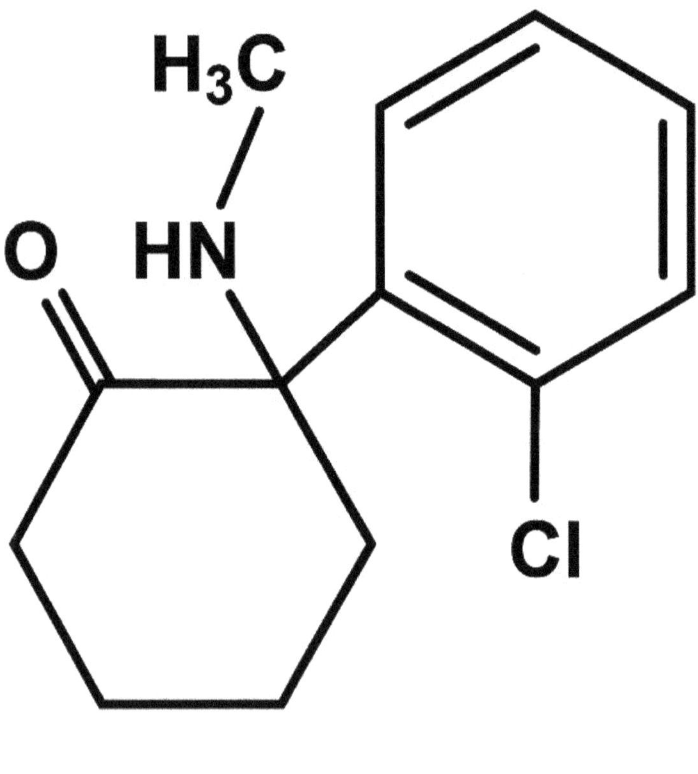

FIGURE 56. **Ketamine.**

As a neuroactive substance Ketamine has been described as a disassociative anaesthetic and analgesic. The term 'disassociative' reflects the consumers feeling of detachment from their environment, including the elimination of the sensation of pain. Unlike many narcotics, which can induce drowsiness, reduced respiratory rate and unconsciousness, ketamine maintains good respiratory integrity and acts as a stimulant of the circulatory system rather than a depressant. The feeling of well-being that often accompanies its use has led to its adoption as a remedy for depression. Esketamine, a stereoisomer of ketamine, was approved as an antidepressant in Canada, the United States, and by the European Medicines Agency in 2019.

As a consequence of its established reputation as a mind-altering drug, the discovery of ketamine as a bioactive component of the *Pochonia chlamydosporia* fungus has received much attention, including speculation as to whether the fungus could conceivably act as a source of ketamine for pharmacological as well as pesticide use. Ferreira *et al.* (2020) draw attention to the potentially varied roles played by ketamine following its isolation from *P. chlamydosporia*. However, so far, there is no positive proof that ketamine can be obtained from the fungus in sufficient concentrations for it to be clinically effective. Research initiatives continue to explore the potential for *P. chlamydosporia* and other closely related fungi to produce ketamine in sufficient concentrations for pharmacological use. Meanwhile ketamine remains a manufactured product.

Ketamine was first synthesized and developed for the Parke Davis drug company in 1962; it was derived from the

NMDA[24] receptor agonist phencyclidine and developed as a relatively safe anaesthetic with fewer hallucinogenic effects than phenylcyclidine (phenylcyclohexyl piperidine (PCP)), the drug it was designed to replace. Phenylcyclidine was approved for use in the United States in 1970 and was widely employed as a surgical anaesthetic during the Vietnam War. The ability of ketamine to block the NMDA receptor results in the inhibition of pain transmission by the spinal cord.

Despite its broad approval as an analgesic. anaesthetic and antidepressant subject to national regulations, the illicit use of ketamine caused it to be classified as a schedule III non-narcotic controlled substance in the U.S.A. in 1999. In Canada Ketamine has been listed as a schedule 1 narcotic since 2005 and in the U.K. ketamine was declared a Class B drug in 2014, meaning that it is illegal to buy or sell. Even in cases where medical supervision is invoked, casualties may still occur due to hallucinogenic time/space distortions. It is presumed that the drowning death of American actor Matthew Perry was a secondary consequence of Ketamine use.

The Magpie Mushroom.

Coprinopsis picacea, commonly known as the magpie mushroom, magpie fungus, or magpie inkcap (FIGURE 57.) is a toxic species of fungus currently ascribed to the family Psathyrellaceae (division Basidiomycota; class Agaricomycetes). The magpie inkcap is frequently found in Europe and Australia, but it also exists in North America where it is less common.

In many, but not all, specimens the dark brown/black fruiting body liquifies as the mushroom reaches maturity,

---

[24] N-methyl-D-aspartate (NMDA) is an agonist (mimic) of the neurotransmitter glutamate

producing an inky fluid. This has resulted in the term 'Inkcap' being added to the Magpie name in members of the group where solubilisation of the cap is evident. This liquidation appears to be a product of autodigestion. Specimens that showed this tendency were initially placed in the Coprinaceae family which comprised all of the 'Inky Cap' mushrooms. However, following 21st century advances in phenotyping, several of the type species *Coprinus* were later incorporated into the Agaricaceae family which became synonymous with the Coprinaceae. The family name Psathyrellaceae is now based on the former Coprinaceae subfamily name Psathyrelloideae. To further confuse matters, the type genus *Psathyrella* has now split into several closely related genera consists of species that do not undergo liquidation of the fruiting cap.

The Magpie Mushroom is characterised by a long whitish stalk that may approach 12" in length supporting a bell-shaped cap that can reach 3" in diameter. The cap is initially pale in colour although this surface skin (velum) disintegrates as the mushroom develops, revealing a brown fruiting body that darkens with age The remnants of the original pale skin continue to adhere to the dark brown cap. Pale remnants of the velar skin of the fruiting body adhering to the dark cap give the mushroom a speckled pattern that has been compared to the plumage of a magpie or woodpecker (FIGURE 57.).

FIGURE 57. **The Magpie Mushroom.** *Coprinopsis picacea.* Remnants of the pale-coloured velum adhering to the cap give a mottled effect that has been compared with magpie or woodpecker plumage.

Beefsteak Fungus.

The beefsteak Fungus *Fistulina hepatica*, also known as *Boletus hepaticus* looks very much like a slab of meat lying near the base of trees from all five continents of the globe. It is a bracket fungus that colonises the live or dead wood of a variety of trees including chestnut and oak from North America and Europe. In Australia it may also be found on the bark of Eucalyptus trees where damage to the bark has occurred. The fungus is responsible for 'brown rot' on several of the trees that it invades.

It is difficult to ignore its resemblance to slabs of raw meat, both in terms of colour and its dimensions. Caps are generally between 3-12" in diameter and 0.75-2.5" thick (FIGURE 58). In addition to beefsteak, a likeness to liver has also been widely noted and is responsible for the incorporation of the Greek for liver, *'hepär'* into the species' name. The carnivorous theme is further extended by the common names 'Tongue Fungus' and 'Ox-Tongue Fungus'. Even juice from the fungus is not unlike blood and has a tendency to leave a red stain on wood that has been colonised by the fungus.

Not surprisingly the meaty appearance of the fungus has prompted interest in its edibility and culinary potential. In this regard, however, the results are somewhat disappointing. Although the fungus is edible it has a bitter taste, due in part to the presence of citric acid in the fungus. Nevertheless, consumers are attracted by its texture, and some are prepared to include it in a meal where other ingredients are of sufficiently strong flavour to disguise the sour taste of the fungus.

FIGURE 58. **Beefsteak Fungus** *Fistulina hepatica.* Despite its meaty appearance the fungus has a sour flavour. (Photo by Kean10 - Own work, CC BY-SA 4.0, https://commons.wikimedia.org/w/index.php?curid=43508714)

Ergot Fungus.

Ergot fungi are a group of fungi belonging to the genus Claviceps. They comprise about 50 species of which *Claviceps Africana* and *Claviceps purpurea* (FIGURE 59.) are probably the best known. The fungi grow on cereal species such as ryegrass and sorghum.

Consumption of ergot-contaminated crops by humans and their livestock can lead to several ailments collectively known as ergotism. Convulsive ergotism derives from the ability of ergot alkaloids to modify the action of neurotransmitters such as dopamine, norepinephrine and serotonin causing muscle spasms. The alkaloids responsible for these neurological effects have a variety of names such as ergotamine, ergoclavine and ergometrine, dependent on variations of their multicyclic structure. Their ability to modulate and sometimes mimic the action of neurotransmitters no doubt stems from their shared structural similarities as derivatives of lysergic acid. Structural similarities with lysergic acid are also responsible for hallucinogenic properties of some ergot alkaloids.

FIGURE 59. **Ergot Fungus.** The infestation of numerous grasses and crops may cause serious illness in consumers.

The psychedelic drug Lysergic Acid Diethylamide (LSD) discovered in 1938 is a derivative of *Claviceps purpurea*. Although the psychedelic properties of some ergots have attracted a small number of users, high doses have resulted in potentially harmful hallucinations leading to irrational behaviour, convulsions, and even death. Ergot does not contain LSD but instead contains lysergic acid as well as its precursor ergotamine. Lysergic acid is a precursor for the synthesis of LSD. Their realized and hypothesized medicinal uses have encouraged intensive research since the 1950s culminating in the development of drugs that are both legal (e.g. bromocriptine) and illegal (e.g., LSD). This initiative has also generated extensive knowledge of the enzymes, genetics and diversity of ergot alkaloid biosynthetic pathways. Slight variations in chemical composition of ergot alkaloids have resulted in a broad range of human pharmacological effects associated with accidental *Claviceps* consumption.

In severe cases of ergotism reported as early as the 16th century, there have been reports of vasoconstriction developing into gangrenous conditions leading to loss of limbs, and spontaneous abortion caused by severe uterine contractions. Other adverse symptoms of ergotism have included an intense burning sensation of the skin that came to be known as St. Anthony's (or Holy) Fire, tremors, insomnia, excessive salivation, vomiting, and occasionally death (Wright 2025). The name St. Anthony's fire is derived from '*The Hospital Brothers of St. Anthony*', an order of monks established in 1095 who specialized in treating the burning sensation experienced by ergotism victims with balms containing tranquilizing and restorative plant extracts.

The earliest documented case of ergotism occurred in AD 857 in the Rhine Valley. This was followed by further outbreaks in Scandinavia and in southwest France, where, in 994, a massive epidemic attributed to ergotism caused an estimated

40,000 deaths. In 1596, the German physician Wendelin Thelius was one of the first to attribute ergotism poisoning to ergot infested grain, and this link between ergotism and fungal infection of grain such as rye was elaborated by the French physician Denis Dodart in 1676 who advocated better management of crops prone to contamination by *Claviceps.* Although this has reduced the threat of human ergotism over several decades the disease remains a concern for farmyard animals fed contaminated grain products (Bennett and Klich, 2003).

In 1778, following another large outbreak of ergotism in Sologne that claimed the lives of about 8000 people, the French physician Abbé Jean-Baptiste Tessier pioneered preventative measures including drainage of fields, crop management such as the compulsory cleaning of grain, and the substitution of potatoes and other crops for ergot-contaminated grain. Some other cereals are naturally resistant to ergot. For example, England was not significantly affected by ergotism because its main cereal crop is wheat, which is resistant to ergot fungi.

Multiple cases of human ergotism have been reported from Europe, North America and Asia over the last 500 years, including reports linking convulsive and hallucinogenic symptoms to cases of bewitchment or possession by the devil. These have included a popular but disputed theory that an outbreak of ergotism may have been at the root of the Salem Witch trials in Massachusetts, U.S.A. in the late 17th century.

Although records documenting the early medicinal and spiritual use of ergot tend to be from Medieval times the mythical history may be much older.

The *Eleusinian Mysteries* were religious rites celebrated annually in ancient Greece honouring the myth of the Goddess Demeter whose daughter Persephone was abducted by

Hades, the King of the Underworld. These mythical stories may have their roots in the Mycenian period from 1350 – 1050 BCE but were certainly part of a major festival celebrating the Afterlife during the Hellenic period of Greek culture from 323 BCE onwards. The essence of the myth focusses on the return of Persephone to her mother Demeter, following her journey to the underworld. The Eleusinian Mysteries are a ritualistic celebration of the return of Persephone from the Afterlife and represent a celebration of eternity and rebirth, including a conception of the Afterlife itself.

FIGURE 60. **5th Century BCE carved marble memorial.** This is thought to represent Persephone and her mother Demeter holding mushrooms, used as part of the Eleusian mystery rites.

These rituals, which have been celebrated over a period of two millennia, were originally conceived to enable participants to communicate with the Afterlife, which is seen as 'a Place of Bliss'.

The historic term 'Elysian Fields' has since been used as a direct reference to this mythical Greek paradise. It is widely assumed that these early 'journeys' to the Elysian Fields were aided by the presence of psychedelic ergot in the concoction known as kykeon, which was drunk as part of the Eleusinian Mysteries ritual. Magic mushrooms have also been implicated as contributors to the psychedelic state associated with the Eleusinian Mysteries celebration (see FIGURE 60.) However, the true involvement of ergot and hallucinogenic mushrooms in this ancient celebration remains a topic of much speculation.

Moving on from the hallucinogenic ergot beverages and their contribution to the mythology of ancient Greece, a positive development in recent decades has been an increased initiative to harness the therapeutic properties of ergot alkaloids and other compounds derived from fungi .

However, as is the case with several natural toxins, the history of ergot alkaloids often illustrates a delicate balance between their potentially harmful effects at higher concentrations but remedial uses at much lower doses. This balance is illustrated by the hazard of premature abortion as a historical consequence of ergot ingestion, compared with its use as a childbirth aid at low doses of ergot alkaloids, a therapeutic function recognised by the German physician Adam Lonitzer as early as 1582.

Therapeutic properties of ergot alkaloids have now been recognised for a broad range of maladies, including gastrointestinal, reproductive and neurological disorders, with the prospect that fungal genomes[25] may be manipulated to generate strains capable of producing therapeutic alkaloids with applications for medicine and agriculture (Panaccione

[25] A 'Genome' is a collective term for the complete set of genes present within a cell or organism.

2023, Wright 2025).

As more becomes known about the remedial properties of ergots, extracts of the fungus have been used increasingly in pharmaceutical preparations such as Cafergot, which is remedy for migraine headaches containing caffeine and ergotamine, and ergometrine which is used to induce uterine contractions and to control bleeding after childbirth.

A Monster Fungus?

It is perhaps fitting to conclude with a reference to a lifeform that epitomises the mystery that has surrounded fungi since their discovery several millennia ago. Part of what might be called 'fungal mystique' relates to the inherent difficulties in understanding their lineage and taxonomic relationship(s) with other organisms; a dilemma exemplified by the title of this book. Few objects illustrate this conundrum better than *Prototaxites.*

Prototaxites (FIGURE 61) are towering trunk-like fossils resembling massive fingers pointing skywards, that have been found in many parts of the globe. These massive fossils, often over 20 feet in height and measuring more than 3 feet in diameter, were first discovered by William Logan in 1843 in fossil beds in the Canadian Gaspé Peninsula dating from the late Silurian to late Devonian period 420-370 million years ago. The discovery was acknowledged by Canadian geologist J.W. Dawson who in 1859 mistakenly identified the fossil as that of a conifer tree trunk. This misidentification was responsible for the generic name *Prototaxites* which translates as 'first yew'.

FIGURE 61. **Prototaxites.** Originally thought to be fossilized trees. Later identified as massive fungal fossils, but.now believed to be fossils of a unique extinct life form.

Since the initial discovery of these fossils, they have been a topic of intense speculation including a long-held opinion by some researchers that they may have been the fossils of an enormous fungus. This theory was based on the resemblance of *Prototaxites* to the fruiting body of some fungal species, despite the massive size of these fossils compared with existing fungi.

To date approximately 14 species of Prototaxites have been discovered in West Africa, Western Asia, Western Europe, North America and Australia. Although their sizes vary, all resemble tree trunks in terms of their height and thickness. Despite the great differences in size between

Prototaxites and existing fungal groups there has been a sustained effort over the last thirty years or so find evidence linking Prototaxites with current fungal groups such as the Ascomycota, Basidiomycota and Muromycota. Examination of *Prototaxites* fossils for fungal characteristics has been greatly hampered by their rock-like consistency but, nevertheless this effort persists using a variety of physical and chemical probes.

One such initiative continues to be carried out on well preserved Prototaxites from the early Devonian Rhynie chert sedimentary deposits in Aberdeenshire in Scotland. These studies have included searches for specific chemical compounds that are indicative of fungi. Examples would include the presence of chitin and β-glucans and other chemical biomarkers that are characteristic of fungal cellular tissue. Other investigations have focussed more on the physical characteristics of sections cut from the 'body' of the prototaxites in a manner somewhat resembling slide tissue preparation for histological examination of biological tissue. These fossil sections have been subjected to confocal laser scanning microscopy, a fluorescence technique specifically designed to examine microstructure. A particularly interesting outcome of this study has been that prototaxites do indeed possess a hyphae-like structure resembling a fungal mycelium but that its pattern of branching differs significantly from any living fungi.

The overall conclusions from the chemical and microscopic research on Rhynie chert fossils state that there is no conclusive support for fungal affinity by prototaxites (Loron *et al.* 2026). However, several researchers do not rule out the possibility that prototaxites were members of a now extinct life-form that doesn't conform to any existing fungal lineage. Remarkably this would, therefore, place prototaxites in a separate and distinct kingdom alongside animals, plants, fungi, protists and bacteria. In some respects, this dilemma echoes the uncertainty that has accompanied fungi throughout their history. As ever, insight into competing ideas concerning their affiliation(s) and evolution will require further fossil exploration and the employment of new genetic and chemical technology. As a result of the resurgence of interest in fungi in recent decades there is no doubt that such initiatives will continue to receive robust support!

Favourable Fungi.

However, probably the subjects of most current interest to mycologists are the form and function of bioactive fungi, with particular focus on species having nutritive, curative and narcotic properties. Other areas of interest relate to the positive or negative influences of fungi on agricultural practices resulting from their enhancement or impedance of crop plants and other biota favoured by humans and their livestock.

Recent advances in biochemistry and genetics have enabled us to manipulate the genome(s) of fungi in several ways. The fungal history of most interest to mycologists relates to the curative and nutritive properties of these organisms as well as their narcotic tendances. It is reasonable to assume that this line of enquiry will play a major role as our research into these extraordinary organisms continues to advance.

General Reading.

Fortey, R. (2024). *Close Encounters of the Fungal Kind.* In Pursuit of Remarkable Mushrooms. Harper Collins. Publ. Glasgow. 336pp. ISBN 978-0-0086-3968-6.

Lawrence, S. (2022).*"The Magic of Mushrooms.* Wellbeck Publ. London & Sydney. 208 pp. ISBN 798-1-78739-906-8.

Lim, M. and Shu, Y. (2022). *The Future is Fung.* Thames & Hudson Publ. Australia. 210 pp. ISBN 798-1-780-76760-8.

Boddy, L. and Ashby, A, (2023). *Fungi. Discover the Science and Secrets Behind the World of Mushrooms.* Darling Kindersley (Penguin Random House), London. ISBN: 978-0-2416-1296-5.

Sheldrake, M. (2021). *Entangled Life. How Fungi Make Our Worlds, Change Our Minds and Shape Our Future.* Bodley Head Publ. London.

References.

Bengtson, S., Rasmussen, B., Ivarsson, M., Muhling, J., Broman, C., Marone, F., Stampanoni, M. and Bekker, A. (2017). *Fungus-like mycelial fossils in 2.4-billion-year-old vesicular basalt.* Nat. Ecol.Evol.1:141–141.https://doi.org/10.1038/ s41559-017-0141.

Burk, W. 1983. *Puffball usages among North American Indians.* J. Ethnobiol. 3(1):55–62.

Hall, I.R. (2003). *Edible and Poisonous Mushrooms of the World.* Portland Timber Press. U.S.A. ISBN: 978-0-88192-5869. 371pp.

Edwards, J. (1985). *The Roman Cookery of Apicius.* Ebury Publ. U.K. ISBN-10: 0712610642. 322pp.

Eroglŭ, C., Seçme, M., Atmaca, P., Kaygusuz, O., Gezer, K., Bağcı, G., and Dodurga, Y. (2016). *Extract of Calvatia gigantea inhibits proliferation of A549 human lung cancer cells.* Cytotechnology. 68(5). 2075-2081. doi: 10.1007/s10616-016-9947-4.

Ferreira, S.R., Machado, A.R.T., Furtado, L.F., Gomes, J.H., de S., de Almeda, R.M., Mendes, T., de O., Maciel, V.N., Barbosa, F.S., Carvalho, L.M., Bueno, L.L., Bartholomeu, D.C., de Araújo, Rabelo, E.M.L., de Pádua, R.M., Pimenta, L.P.S. and Fugiwara, R.T. (2020). *Ketamine can be produced by Ponchonia chlamydosporia: an old molecule and a new helmintic?* Parasites and Vectors. https://doi.org/10.1186/s13071-020-04402-w

Fukasawa, K., Hamano, K., Kaga, K., Akai, D., and Takehi, T. (2024). *Spatial resources arrangement influences both network structures and activity of fungal mycela: A form of pattern recognition?* Fungal Ecol. 72, 101387.

Guzmán, G., Allen, J.W., and Gartz, J. (1998). *A Worldwide Geographic Distribution of the Neurotropic Fungi, an Analysis and Discussion*. Annali del Musco Civico di Roverto (14) 189-280.

Hall, I.R. Stevenson, S.L., Buchanan, P.K., Yun, W. and Cole, A.L.J. (2003). *Edible and Poisonous Mushrooms of the World.* Portland: Timber Press. ISBN 978-0-88192-586-9. 372pp.

Hirt, R.P., Logsdon, J.M. Jr., Healy, B. and Embly, T.M. (1999). *Microsporidia are related to Fungi: Evidence from the largest subunit of RNA polymerase II and other proteins.* Proc. Natl. Acad. Sci. USA. (96) 580-585.

Hoffman, D., Preuss, G. and Mëtzler, C. (2015). *Evidence for biological shaping of hair ice*. Biogeosciences 12(14) 4761-4273. doi:10.5194/bg-12-4261-2015.

Khajuria, A. and Nonzom, S. (2024). *Fungi as a Potential Source of Polyketides. A Review*. Appl. Biochem. Microbiol. 60, pp.554-598. https://doi.org/10.1134/S0003683823603451.

McKenna, T. K. (1992). *Food of the Gods. A Radical History of Plants, Psychedelics and Human Evolution*. Random House. ISBN 9780553371307. 336pp.

Money, N.P. (2024). *Molds, Mushrooms, and Medicines: Our Lifelong Relationship with Fungi*. Princeton University Press, Princeton, New Jersey, U.S.A. ISBN 9780691236308. 240pp.

Mori, K., Obara, Y., Hirota, M., Azumi, Y., Kinugasa, S., Inatomi, S. and Nakahata, N. (2008). *Nerve Growth Factor-Inducing Activity of* ***Hericium erinaceous*** *1321N1 Human Astrocytoma cells*. Biological and Pharmaceutical Bulletin 31(9), pp. 1727-1732.

Panaccione, D.G. (2023). *Derivation of the multiply-branched ergot alkaloid pathway of fungi*. Microbial Biotechnology. 16, pp 742-756. Doi 10. 1111/1751. 14214.

Shimbo, M., Kawagishi, H. and Yokogoshi, H. (2005). *Erinacine A Increases Catecholamine and Nerve Growth Factor Content in the Central Nervous System of Rats.* Nutrition Research 25(6), pp. 617-623.

Ventenat, E. P. (1798). *Dissertation sur le genre Phallus* (M. Kuo, Trans.). *Mémoires de l'Institut des Sciences et Arts* 1: 517–518.

Wasson, R.G. (2021). *Soma. Divine Mushroom of Immortality*. Sungazer Press. ISBN-10 156838001. 406pp.

Wright, D.A. (2025). *Natural Toxins. The Good, The Bad and The Deadly.* Llyfrau Cambria Books, Wales, U.K. ISBN 978-1-0685688-3-1.

Other books in this series by David Wright available from Cambria Publishing.

Print ISBN 978-1-0685688-3-1

222 pages

www.ingramcontent.com/pod-product-compliance
Lightning Source LLC
LaVergne TN
LVHW050913210926
848482LV00012B/426
* 9 7 8 1 9 1 9 2 6 9 8 9 4 *